Urban Water Systems
& Floods IV

WITPRESS

WIT Press publishes leading books in Science and Technology.
Visit our website for the current list of titles.
www.witpress.com

WITeLibrary

Home of the Transactions of the Wessex Institute.
Papers published in this volume are archived in the WIT eLibrary in volume 208 of WIT Transactions on
the Built Environment (ISSN 1743-3509). The WIT eLibrary provides the international scientific
community with immediate and permanent access to individual papers presented at WIT conferences.
http://library.witpress.com.

EIGHTH INTERNATIONAL CONFERENCE ON
FLOOD AND URBAN WATER MANAGEMENT

FRIAR 2022

CONFERENCE CHAIRMEN

S. Mambretti
Polytechnic of Milan, Italy
Member of WIT Board of Directors

D. Proverbs
University of Wolverhampton, UK

INTERNATIONAL SCIENTIFIC ADVISORY COMMITTEE

ORGANISED BY

Wessex Institute, UK
University of Wolverhampton, UK
Polytechnic of Milan, Italy

SPONSORED BY

WIT Transactions on the Built Environment
International Journal of Environmental Impacts

WIT Transactions

Wessex Institute
Ashurst Lodge, Ashurst
Southampton SO40 7AA, UK

We would like to express thanks to all the conference Chairs and members of the International Scientific Advisory Committees for their efforts during the 2022 conference season.

Conference Chairs

Joanna Barnes
University of the West of England, UK

Juan Casares
University of Santiago de Compostela, Spain
(Member of WIT Board of Directors)

Alexander Cheng
University of Mississippi, USA
(Member of WIT Board of Directors)

Pilar Chias
University of Alcala, Spain

Pablo Diaz Rodriguez
University of La Laguna, Spain

Andrea Fabbri
University of Milano-Bicocca, Italy

Fabio Garzia
University of Rome "La Sapienza", Italy

Massimo Guarascio
University of Rome "La Sapienza", Italy

Santiago Hernandez
University of A Coruna, Spain
(Member of WIT Board of Directors)

Massimiliano Lega
University of Naples Parthenope, Italy

Mara Lombardi
University of Rome "La Sapienza", Italy

James Longhurst
University of the West of England, UK

Elena Magaril
Ural Federal University, Russia

Stefano Mambretti
Polytechnic of Milan, Italy
(Member of WIT Board of Directors)

Jose Manuel Mera
Polytechnic University of Madrid, Spain

Jose Luis Miralles i Garcia
Polytechnic University of Valencia, Spain

Giorgio Passerini
Polytechnic University of Le Marche, Italy
(Member of WIT Board of Directors)

David Proverbs
University of Wolverhampton, UK

Elena Rada
Insubria University, Italy

Stefano Ricci
University of Rome, La Sapienza

Graham Schleyer
University of Liverpool, UK

Stavros Syngellakis
Wessex Institute, UK
(Member of WIT Board of Directors)

International Scientific Advisory Committee Members 2022

Borna Abramovic University of Zagreb, Croatia

Tawfiq Abuhantash American University of Ras Al Khaimah, UAE

Alejandro Acosta Collazo Autonomous University of Aguascalientes, Mexico

Khalid Al Saud King Saud University, Saudi Arabia

Ghassan Al-Dweik Palestine Polytechnic University, Palestine

Hind Algahtani Imam Abdulrahman bin Faisal University, Saudi Arabia

Abdulkader Algilani King Abdulaziz University, Saudi Arabia

Mir Ali University of Illinois at Urbana-Champaign, USA

Bakari Aliyu Taraba State University, Nigeria

Samar Aljahdali University of Jeddah, Saudi Arabia

Hussain Al-Kayiem Universiti Teknologi PETRONAS, Malaysia

Jose Ignacio Alonso Polytechnic University of Madrid, Spain

Reem Alsabban University of Jeddah, Saudi Arabia

Sultan Al-Salem KISR, Kuwait

Andrea Antonucci University of Roma 3, Italy

Srazali Aripin International Islamic University Malaysia, Malaysia

Eman Assi American University of Ras Al Khaimah, UAE

Sahar Attia Cairo University, Egypt

Jihad Awad Ajman University, UAE

Warren Axelrod C. Warren Axelrod LLC, USA

Mohammed Bagader Umm Al-Qura University, Saudi Arabia

Azizi Bahauddin Universiti Sains Malaysia, Malaysia

Francine Baker Wolfson College, UK

Marco Baldi Marche Polytechnic University, Italy

Michael Barber University of Utah, USA

Socrates Basbas Aristotle University of Thessaloniki, Greece

Joao Batista University of São Paulo, Brazil

Gianfranco Becciu Politecnico di Milano, Italy

Michael Beer Leibniz Universitat Hannover, Germany

Khadija Benis c5Lab, Portugal

Marco Bietresato Free University of Bolzano, Italy

Alma Bojorquez-Vargas Autonomous University of San Luis Potosí, Mexico

Daniel Bonotto UNESP, Brazil

Colin Booth University of the West of England, UK

Carlos Borrego University of Aveiro, Portugal

Bouzid Boudiaf Ajman University, UAE

Zuzana Boukalova VODNÍ ZDROJE, a.s., Czech Republic

Djamel Boussaa Qatar University, Qatar

Roman Brandtweiner Vienna University of Economics and Business, Austria

Roger Brewster Bond University, Australia

Andre Buchau University of Stuttgart, Germany

Raul Campos RCQ Structural Engineering, Chile

Richard Carranza Carranza Consulting, USA

Paul Carrion Mero ESPOL Polytechnic University, Ecuador

Joao-Manuel Carvalho Universidade de Lisboa, Portugal

Ana Cristina Paixao Casaca Isel, Portugal

Ricardo Castedo Universidad Politécnica de Madrid, Spain

Robert Cerny Czech Technical University Prague, Czech Republic

Camilo Cerro American University of Sharjah, UAE

Vicent Esteban Chapapria Polytechnic University of Valencia, Spain

Galina Chebotareva Ural Federal University, Russia

Hai-Bo Chen University of Science & Technology of China, China

Jeng-Tzong Chen National Taiwan Ocean University, Taiwan

Weiqiu Chen Zhejiang University, China

Rémy Chevrier SNCF Innovation & Research, France

Marco Schiavon University Of Trento, Italy

Michal Sejnoha Czech Technical University, Czech Republic

Marichela Sepe University of Naples Federico II, Italy

Angela Baeza Serrano Global Omnium Medioambiente, S.L., Spain

Leticia Serrano-Estrada University of Alicante, Spain

Wael Shaheen Palestine Polytechnic University, Palestine

Viktor Silbermann Fichtner GmbH & Co KG, Germany

Luis Simoes University of Coimbra, Portugal

Nuno Simoes University of Coimbra, Portugal

Sradhanjali Singh CSIR-NEERI, Delhi, India

Rolf Sjoblom Luleå University of Technology, Sweden

Leopold Skerget Slovenian Academy of Engineering, Slovenia

Vladimir Sladek Slovak Academy of Sciences, Slovakia

Alexander Slobodov ITMO University, Russia

Lauren Stewart Georgia Institute of Technology, USA

Elena Strelnikova National Academy of Sciences of Ukraine, Ukraine

Michel Olivier Sturtzer French-German Research Institute of Saint-Louis, France

Miroslav Sykora Czech Technical University of Prague, Czech Republic

Antonio Tadeu University of Coimbra, Portugal

Paulo Roberto Armanini Tagliani Federal University of Rio Grande, Brazil

Hitoshi Takagi Tokushima University, Japan

Kenichi Takemura Kanagawa University, Japan

Manal Yehia Tawfik El Shourok Academy, Egypt

Filipe Teixeira-Dias University of Edinburgh, UK

Roberta Teta University of Naples Federico II, Italy

Norio Tomii Nihon University, Japan

Juan Carlos Pomares Torres University of Alicante, Spain

Vincenzo Torretta Insubria University, Italy

Sophie Trelat IRSN, France

Carlo Trozzi Teche Consulting srl, Italy

Sirma Turgut Yildiz Technical University, Turkey

Charles Tushabomwe-Kazooba Mbarara University of Science & Technology, Uganda

Elen Twrdy University of Ljubljana, Slovenia

Maria Valles-Panells Polytechnic University of Valencia, Spain

Thierry Vanelslander University of Antwerp, Belgium

Baxter Vieux Vieux & Associates, Inc., USA

Antonino Vitetta Mediterranea University of Reggio Calabria, Italy

Jaap Vleugel Delft University of Technology, Netherlands

Giuliano Vox University of Bari, Italy

Adam Weintritt Gdynia Maritime University, Poland

Alvyn Williams Soft Loud House Architects, Australia

Ben Williams University of the West of England, UK

Shahla Wunderlich Montclair State University, USA

Wolf Yeigh University of Washington, USA

Victor Yepes Universitat Politecnica de Valencia, Spain

Montserrat Zamorano University of Granada, Spain

Giuseppe Zappala National Research Council, Italy

Chuanzeng Zhang University of Siegen, Germany

Dichuan Zhang Nazarbayev University, Kazakhstan

Urban Water Systems & Floods IV

Editors

S. Mambretti
Polytechnic of Milan, Italy
Member of WIT Board of Directors

D. Proverbs
University of Wolverhampton, UK

WITPRESS Southampton, Boston

Editors:

S. Mambretti
Polytechnic of Milan, Italy
Member of WIT Board of Directors

D. Proverbs
University of Wolverhampton, UK

Published by

WIT Press
Ashurst Lodge, Ashurst, Southampton, SO40 7AA, UK
Tel: 44 (0) 238 029 3223; Fax: 44 (0) 238 029 2853
E-Mail: witpress@witpress.com
http://www.witpress.com

For USA, Canada and Mexico

Computational Mechanics International Inc
25 Bridge Street, Billerica, MA 01821, USA
Tel: 978 667 5841; Fax: 978 667 7582
E-Mail: infousa@witpress.com
http://www.witpress.com

British Library Cataloguing-in-Publication Data

A Catalogue record for this book is available
from the British Library

ISBN: 978-1-78466-469-5
eISBN: 978-1-78466-470-1
ISSN: 1746-4498 (print)
ISSN: 1743-3509 (on-line)

The texts of the papers in this volume were set individually by the authors or under their supervision. Only minor corrections to the text may have been carried out by the publisher.

Preface

This volume contains papers presented at the 8th International Conference on Flood and Urban Water Management, which, scheduled in Milano, Italy, was held on line due to the Coronavirus pandemic. The Conference was organised by the Wessex Institute, in collaboration with the Politecnico di Milano in Italy and University of Wolverhampton in the UK.

This is the eighth conference of this successful series. The conference started at the Institution of Civil Engineers in London in 2008 and was reconvened in Milano in 2010, Dubrovnik in 2012, Poznan in 2014, Venice in 2016, A Coruña in 2018 and online in 2020 due to the Covid-19 pandemic. Since 2012 a parallel seminar on the Design, Construction, Maintenance, Monitoring and Control of Urban Water has taken place which is now an integral part of the conference.

Flooding is a global phenomenon that claims numerous lives worldwide each year. The primary effects of flooding include loss of life and damage to buildings and other structures, including bridges, sewerage systems, roadways and canals. Floods also frequently damage power transmission and sometimes power generation, the effects of which include loss of drinking water treatment and water supply, which may result in loss of drinking water or severe water contamination. It may also cause the loss of sewage disposal facilities.

The increased frequency of flooding in the last few years, coupled with climate change predictions and urban development, suggest that these impacts are set to worsen in the future. How we respond, and importantly adapt, to these challenges is key to developing our long-term resilience at the property, community and city scale.

Planning for flood safety involves many aspects of analysis and engineering, including observation of previous and present flood heights and inundated areas, statistical, hydrologic, and hydraulic model analyses, mapping inundated areas and flood heights for future flood scenarios, long-term land use planning and regulation, engineering design and construction of structures to control or withstand flooding, intermediate-term monitoring, forecasting, and emergency-response planning, and short-term monitoring, warning and response operations.

As our cities continue to expand, their urban infrastructures need to be re-evaluated and adapted to new requirements related to the increase in population and the growing areas under urbanization. This conference also considers these problems and deals with two main urban water topics: water supply systems and urban drainage.

Topics such as contamination and pollution discharges in urban water bodies, as well as the monitoring of water recycling systems are currently receiving a great deal of attention from researchers and professional engineers working in the water industry. Water distribution networks often suffer substantial losses, which represent wastage of energy and treatment. Effective, efficient and energy saving management is necessary in order to optimize their performance. Sewer systems are under constant pressure due to growing urbanization and climate change, and the environmental impact caused by urban drainage overflows is related to both water quantity and water quality.

FRIAR seeks to develop an improved understanding of emerging flood risk management and urban water management challenges, drawing on the expertise of numerous disciplines and considering a range of responses. The conference provides a rich forum for the development of innovative solutions that can help bring about multiple benefits toward achieving integrated flood risk and urban water management strategies and policy.

The meeting attracts researchers, academics and practitioners actively involved in improving our understanding of urban water systems and flood events. It brings together social scientists, surveyors, engineers, scientists, and other professionals from many countries involved in research and development activities in a wide range of technical and management topics related to urban water and flooding and its impacts on communities, property and people.

The papers included in this volume make an improved understanding of emerging flood risk management and urban water management challenges, drawing on the expertise of numerous disciplines and considering a range of responses. The conference provided a rich forum for the development of innovative solutions that can help bring about multiple benefits toward achieving integrated flood risk and urban water management strategies and policy.

These papers, like others presented at Wessex Institute conferences, are referenced by CrossRef and appear regularly in suitable reviews, publications and databases, including referencing and abstracting services. They are also archived online in the WIT eLibrary (http://www.witpress.com/elibrary) where they are permanently available in Open Access format to the international scientific community.

The Editors would like to thank the authors for their contributions, as well as the member of the International Scientific Advisory Community of the Conference for their invaluable help in reviewing the papers.

The Editors, 2022

Contents

SECTION 1
FLOOD MODELLING

WHAT HAPPENED IN 2021? ANALYZING THE BIGGEST NEGRO RIVER FLOOD IN MANAUS, BRAZIL

JUSSARA SOCORRO CURY MACIEL[1,2], LUNA GRIPP SIMÕES ALVES[1],
BERNARDO FERREIRA OLIVEIRA[1], RENATO CRUZ SENNA[3]
& VINICIUS DOS SANTOS ALBUQUERQUE[4]
[1]Geological Survey of Brazil, Brazil
[2]Federal Center for Technological Education of Amazonas, Brazil
[3]National Institute of Amazonian Research of Brazil, Brazil
[4]State University of Amazonas, Brazil

ABSTRACT

Flood and ebb processes are common events in any hydrological system. In some cases, due to natural or anthropogenic conditions, such events can take place in an extreme manner, causing a lot of damage to the population. In 2021, in several municipalities in the Amazon basin, rivers reached levels higher than the maximum observed until then, making this year the biggest flood in the entire history of monitoring. Most Amazonian rivers have a high annual pulse of floods, as a result of the precipitation period in the upper part of their large basins. Most of the floodplains located in the central region of Amazonia become inundated from May to July, from which the water drains into the river systems slowly over the drought time. In the Amazon Basin, extreme events are mainly related to El Niño or La Niña events, resulting in some big floods and a long rainfall period. In 2021, the Negro river level exceeded the maximum level observed in the entire 119-year historical series of monitoring. On 30 May 2021, the previous record of 29.97 m observed in 2012 was equaled. The river continued to rise until reaching the level of 30.02 m on 16 June 2021. Other stations monitored by Geological Survey of Brazil, which were accomplished through bulletins, reached historical records in the same year, such as São Gabriel da Cachoeira, Barcelos and Manaus (Negro river), Manacapuru (Solimões river), Careiro da Várzea (Amazon river basin), Itacoatiara and Parintins (Amazon river), all located in the state of Amazonas. This study analyzes the conditions that favored the event of the greatest flood recorded in the Rio Negro in 2021. Some factors contribute to the flooding event, such as the rainfall regime distributed throughout the basin and how the main river and its various tributaries behave during the flooding period.

Keywords: flood record, Amazon basin, Negro river.

1 INTRODUCTION

The 2021 flood event represented the largest and most impactful flood event in its entire history of hydrological monitoring for the state of Amazonas, Brazil. 57 of 62 cities in Amazonas had recognized an emergency situation. In the state capital, Manaus, where data on river levels have been recorded since 1902, the level of the Rio Negro surpassed all previous records, confirming 2021 as the largest flood in the last 119 years in the region. Extreme flood events occurred mainly in the "central" region of the Amazon basin, including Manaus, Manacapuru (Solimões river), Careiro da Várzea, Itacoatiara and Parintins (Amazon river). Extreme flooding was also observed along the entire Rio Negro, such as São Gabriel da Cachoeira and Barcelos cities also reaching river levels never seen before.

The floods that occur on the edge of Manaus and its surroundings are due, for the most part, to the contributions of the Solimões and Negro rivers. They are floods that have a long journey time, due to the gigantic size of the hydrographic basin and the small slope observed in the beds of its main bodies of water [1]. The Amazon's many tributaries do respond in a similar fashion, but when all these inputs are integrated together down the main stem of the river, this results in a single monomodal flood wave that occurs with regularity on an annual basis. The flood wave elevation range at Manaus Station is around 9.5 m, with high water in June–July and the low water in October–November [2].

In the Amazon River Basin, backwater effects regulate the flow dynamics in the downstream reaches of main rivers. As examples of these effects in the basin, there is the influence of the main Amazon River on its tributaries' water levels [3], [4]. Hydrological regimes within the basin vary significantly. The peak water discharge of the Solimoes/Amazonas, Madeira and Negro Rivers are offset in time. For the upper Negro River, the high-water period arrives in the second half of the year [5].

Negro river level in Manaus is verified by monitoring some stations located before and in different waterbodies that contribute to Manaus station. These stations have scale meters installed on the banks and are accompanied by local people daily. The record of rivers rise and fall in these seasons are recorded in a weekly newsletter and published on the institutional homepage (www.cprm.gov.br/sace) of the monitoring responsible sector and it has a great importance to the resident population, the state and city halls, as well as for academic researchers [6].

The maximum rainfall in the Southern most parts of the Amazon River basin usually occurs in December, January and February. The maximum rainfall in the central basin along the Solimões–Amazon mainstem is in February, March and April and six months earlier than maximum rainfall in the northernmost parts of the basin (June–July–August) [4]. Daily water levels at the Manaus Port records since 1903 assists the Geological Survey of Brazil (CPRM) to identify severe droughts (15.80 m) and floods (29.00 m) in Manaus and characterize their frequency, duration, and severity. These river levels are critical for the functioning of the port and are used to declare emergency status in the city [8]. Meade et al. [4] present a comparison of the stage hydrograph for Manaus at the bottom with the hydrographs for station on the upper Negro River in the upper and with the hydrograph for the mainstream Solimões River at Manacapuru, shows that stages in the downstream reaches of the Negro reflect the stage of the mainstem.

In this paper we present the aspects that could be the responsible for the biggest flood in Negro river through more than a hundred years of observation and offer a kind of operational instrument to future studies and analysis.

2 CLIMATE

Hydrological drought and flood extreme events were quantitatively defined to occur when daily water levels in Manaus fall below 15.80 m or rise above 29.00 m. In the Solimões–Amazonas system, the flood intensity depends on the rainfall regime of the entire basin. Rainfall decreases in the Amazon are partially associated with the phenomenon popularly known as "El Niño" that produces severe drought or ebb and "La Niña" cause flooding intense [6].

Trigg et al. [2] mention that rainfall in the Amazon has a pronounced pattern across the Amazon basin, linked to global climate processes. The Intertropical Convergence Zone (ITCZ), where winds converge from the southern and northern hemispheres, induces wet and dry periods alternately in the northern and southern sides of the basin. South of the equator there is a wet period from December to February and north of the equator the dry period is from June to August. The South Atlantic Convergence Zone (SACZ), another axis of convergent winds oriented northwest southeast across southeast Brazil and into the southwest Atlantic Ocean also increases rainfall in these areas (see Fig. 1). Such combined rainfall patterns maintain a high base flow in the Amazon River, which added to the water storage in the floodplain, contributes to each year's flood event.

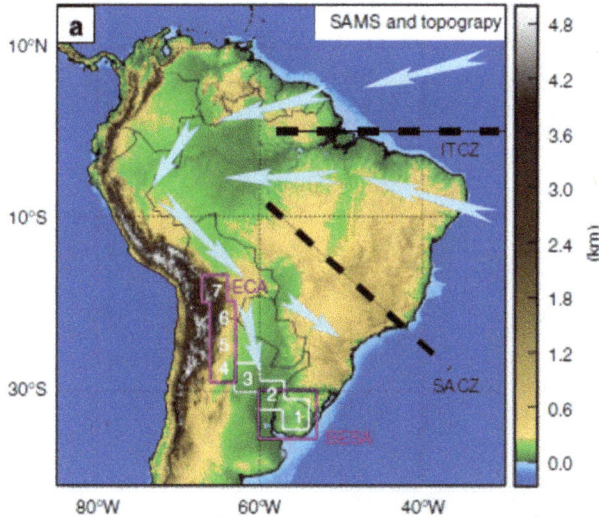

Figure 1: Geographic and climatic setting. Topography and simplified South American Monsoon System mechanisms. The boxes labelled 1 to 7 indicate the climatological propagation path of extreme events [10].

Large Amazonian rivers tend to have annual, predictable high-amplitude flood pulses, due to the seasonality of precipitation in their large catchments. Most Amazonian floodplains become inundated from January to March from which the water drains into the river systems slowly over the following months. Most of the rivers in the Amazon network reach flood levels during May to July and low levels during September to November, with variations in the average timing of peak/low discharge across the Amazon basin. The large drainage basins integrate precipitation variability; the river valley topography and wetlands attenuate and delay the flood wave. In the Central Amazon region, the maximum water level coincides with the beginning of the dry season, as the water needs 2–3 months to flow the several hundred kilometers from the headwaters to this region. Thus, interannual variability in the maximum water levels, in these free-flowing rivers, results from rainfall variability over the catchment regions in the months prior to the peak water level. Such regularity and temporal predictability enables statistical seasonal forecast models to predict the magnitude of hydrological peak water levels, which have high interannual variation [9].

Between December 2020 and January 2021, large areas of the western Amazon basin, a region that drains to Manaus and areas nearby, presented rainfall volumes above the climatology normally observed in the period (Fig. 2).

In February 2021, above-average rains were concentrated in the state of Acre, causing severe flooding in the state's municipalities, and also in cities in Amazonas close to the region, such as Ipixuna, Boca do Acre and Envira. In March, rainfall classified as "very rainy" to "extremely rainy" was observed throughout the entire western Amazon. In April, the entire Negro River basin maintained this pattern, as well as the entire region close to the Solimões main channel, and some points further south of the basin. In May, a large part of the basin presented rainfall below expectations for the period. In June, the signs were less intense, with regions scattered across the basin ranging from "trend to very dry" to "trend to very rainy".

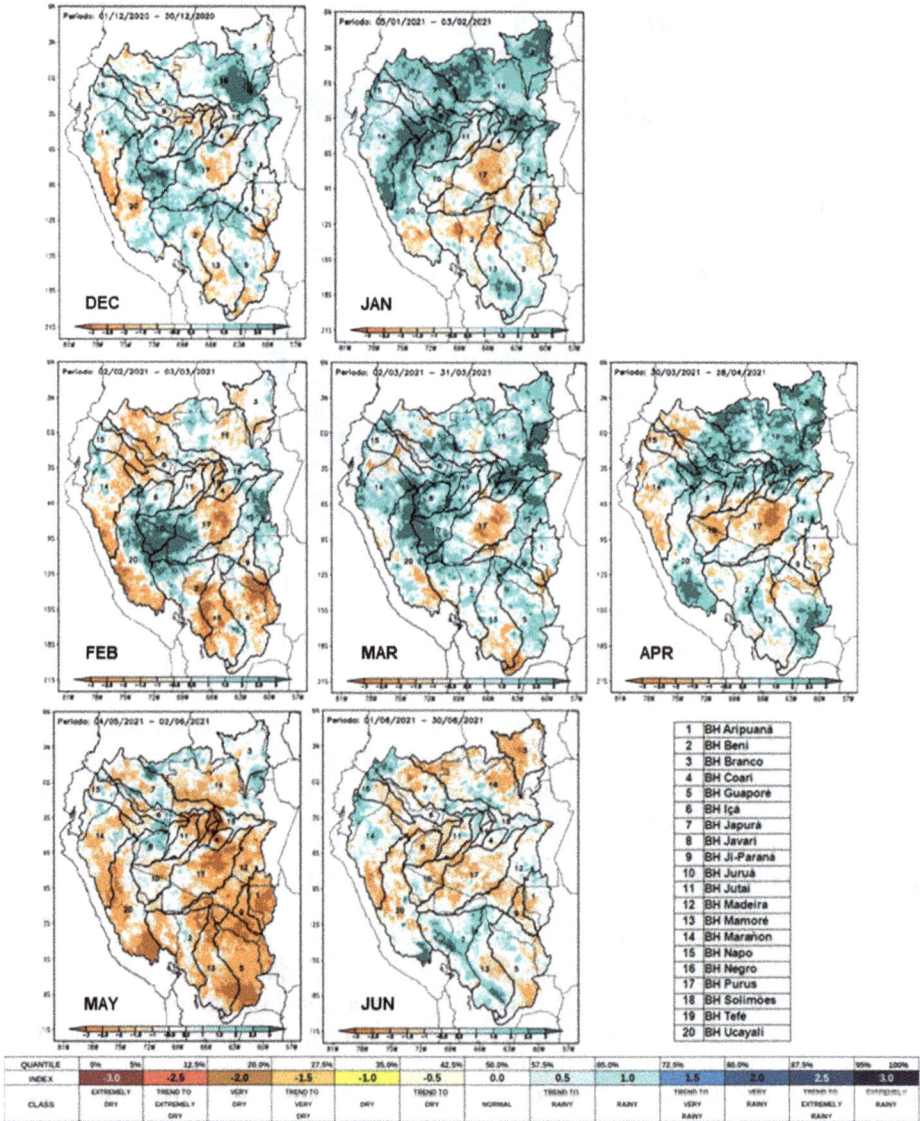

Figure 2:   Monthly accumulated precipitation anomalies distribution between December 2020 and June 2021. *(Source: http://ftp.cptec.inpe.br/modelos/io/produtos/ MERGE/.)*

Espinoza et al. [11] reveal that extreme flood years (1986, 1993, 1999 and 2012) are characterized by negative sea surface temperature (SST) anomalies on the central equatorial Pacific during austral summer and spring. This temperature alteration generates geopotential height anomalies over the subtropical South Atlantic and South America that contributes to a southern humidity flux anomaly over the Amazon. The climatological event "La Niña" shows a strong correlation with an increase in precipitation volumes in the region. As of the second half of 2020, the index that characterizes the event ("Oceanic Niño Index, ONI") on the region called Nino 3.4 began to show signs of its establishment. Table 1 shows the monthly ONI values observed in 2020 and 2021, with the characteristic quarters of the phenomenon being marked in blue. From the third consecutive quarter of indices in blue, in the case at the end of the September–October–November 2020 quarter, the La Niña event was officially recognized.

Table 1:   Oceanic Niño Index (ONI) in Niño 3.4 region. *(Source: https://origin.cpc.ncep. noaa.gov/products/analysis_monitoring/ensostuff/ONI_v5.php.)*

	DJF	JFM	FMA	MAM	AMJ	MJJ	JJA	JAS	ASO	SON	OND	NDJ
2020	0.5	0.5	0.4	0.2	−0.1	−0.3	−0.4	−0.6	−0.9	−1.2	−1.3	−1.2
2021	−1.0	−0.9	−0.8	−0.7	−0.5							

The changes in the Amazonian rainfall patterns have been related to the intensification of the Hadley and Walker circulations. It was estimated a strengthened of the Hadley circulation by 1–5% in the period of 1979–2009. Substantial warming of the tropical Atlantic and simultaneously cooling of the tropical Pacific since the 1990s plays a central role in rainfall trends [12]. From the establishment of La Niña, an influence on atmospheric circulation through the Walker Cell is expected, whose main characteristic is the increase in convection and, consequently, greater cloud development and increased precipitation in different regions of the world. Over the Western Amazon, increases in precipitation are usually observed over the north of the region during the rainy season of the region (November to May), as occurred in the year 2021. These weather patterns, observed since the end of 2020, determined the great magnitude of the flood event observed in 2021, which was configured as the biggest flood in the entire history of hydrological monitoring in the state of Amazonas.

3 RIVER LEVEL

The city of Manaus, even located in the Negro River basin, is greatly influenced by the Solimões–Amazonas system, through the hydraulic backwater effect (Fig. 3). The Solimões River is the main source of the Amazon River, the behavior of the Solimões River in its more downstream stretch represents well what occurs along the westernmost stretch of the Amazon River.

In average, Negro river in Manaus takes 234 days of rising and 130 days of descent. The climb is smooth and striking, which facilitates predictability. Negro river levels have been measured since September 1902, with a historical series of 119 years. Analyzing this historical series, 6% of the floods occurred in May, 75% in June and 19% in July. The Manaus gage provides an accurate, long-term record of stage; it has been maintained at the same site by Manaus Harbor over the period of record [6].

For example, in 2009, the second highest flood at Manaus Station, not only Manaus registered a flood record in the first half of the year. At least five other locations also recorded maximum indices of their historical series and they could be presented as Manacapuru, Itapéua, Parintins, Careiro and the community called Forte de Nossa Senhora das Graças, situated in the gutter of Juruá [13]. In 2012, the flow in January was already with great measurements and this for both gutters (Negro and Solimões). In the comparison of flows, in Manacapuru, the discharges obtained in the months preceding the flood were higher than those recorded in before years. For instance, Manacapuru discharge in 2011 was 63,139 m^3/s, however in 2012 it was 105,641 m^3/s, both in January [6].

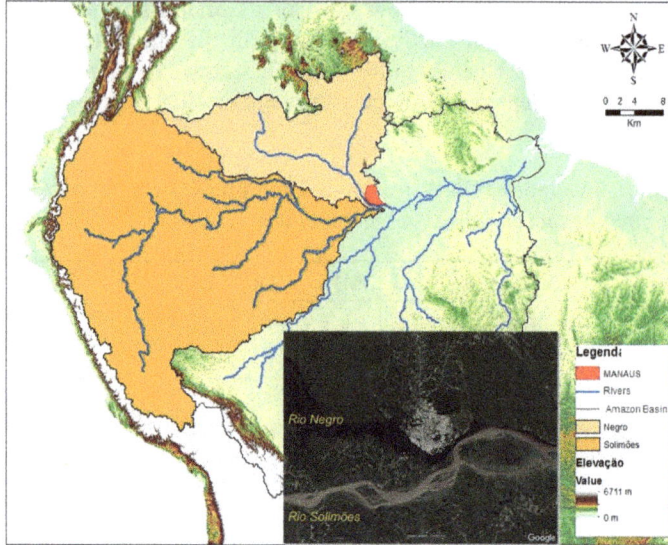

Figure 3: Manaus location in relation the main basins (Negro and Solimões).

In 2021, for the Rio Negro basin, the large volumes precipitated in December made the river level start the year with levels higher than expected. In February, normal rainfall in the drainage basin allowed for a drop in the river level. However, from March onwards, the river began to rise again, always at levels above expectations. In São Gabriel da Cachoeira, the maximum level previously observed was surpassed in May, continuing to rise to the level of 12.68 m (on 6 May 2021), 51 cm above the previous maximum. In Barcelos, the historic record was also surpassed, reaching the maximum that year (10.46 m) on 27 June 2021, also determining the largest flood of the historic series in the region. In Santa Isabel do Rio Negro (Tauruquara station), the same patterns were observed, but the historical record was not surpassed (see Fig. 4).

The pattern observed throughout 2021 at the monitored stations in the Amazon basin (Careiro da Várzea, Itacoatiara and Parintins) is similar to that observed in Manacapuru, in the Solimões river. In all seasons, river levels rose significantly between January and February 2021, and maintained an average rising pattern, which caused a river level behavior with levels significantly higher than expected over the following months.

Thus, the behavior of the river at Manaus station is similar to that observed in Manacapuru, Itacoatiara and Parintins, for example (Fig. 5). Between the months of January and February, all the excessive rain that had been precipitated in the previous months led to a significant increase in the level of the river. Since then, the combination of precipitation patterns observed in the basins that drain into the region caused, in general, a rate of rise in level considered normal over the following months. Therefore, a beginning of the year with levels higher, associated with an average subsequent ascent speed, culminated in a river level significantly higher than expected at the time when the river annually reaches its maximum level (June and July). Consequently, in 2021, Negro river level exceeded the maximum level previously observed in the entire 119-year historical series of station data. On 30 May 2021, the previous record of 29.97 m observed in 2012 was equaled. The river continued to rise until reaching the level of 30.02 m on 16 June 2021.

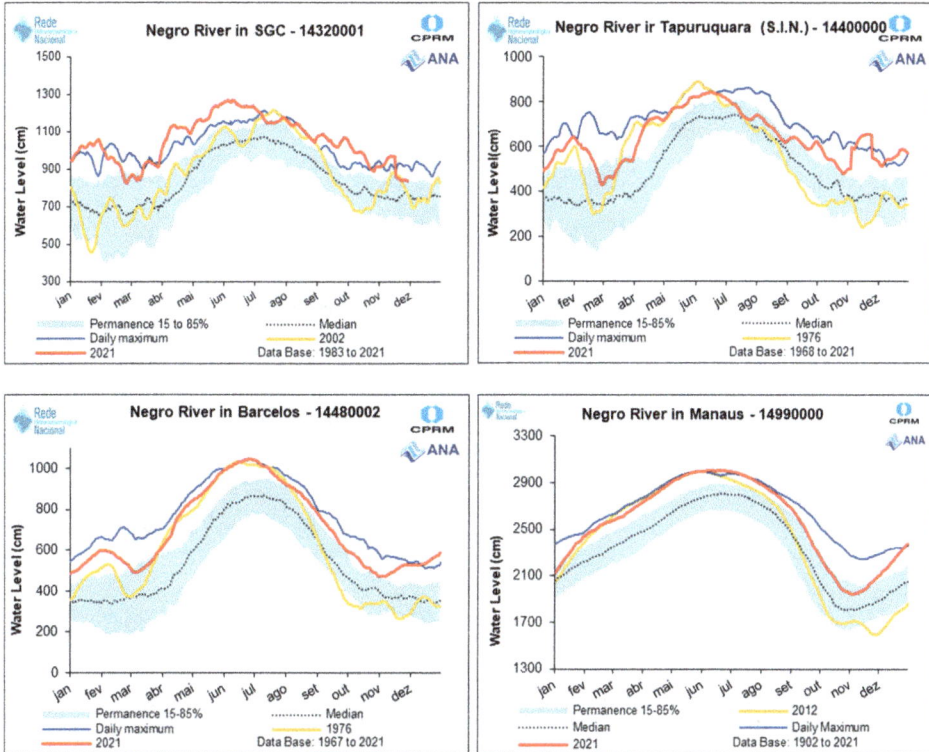

Figure 4:   Negro river stations graphs during 2021. The yellow lines represents the year that the maximum level was observed before 2021.

In addition to the great magnitude in terms of impacts associated with the flooding observed that year, the event was also very expressive in terms of temporal duration, as several parts of the municipality were flooded for a long time. The flood stage, which for Manaus is considered to be 27.50 m, when the first point of its urban area begins to be flooded, was reached on 4 April 2021. The severe flood stage, however, was 29.00 m, when the central region of the municipality begins to be flooded, was reached on 30 April 2021. As the river rose much above these levels, even after the beginning of the ebb process, it took many days for the river to leave its present stage above the reference ones. The severe flood stage, for example, was yet observed until the end of July. In other words, the municipality of Manaus spent 90 days in severe flooding in 2021. In 2012, the year of the previous record flood, the river had remained approximately 75 days in this situation.

4 CPRM FORECAST

Since 1989, the Geological Survey of Brazil in Manaus has been developing the "Amazon Hydrological Alert System" where the annual flood and ebb monitoring process is performed in the Solimões–Negro–Amazonas system. Among the products generated by the project is the Manaus Flood Alert, which presents the forecast of the maximum stage to be reached by the Negro River in Manaus each year. The results are released to the relevant agencies and the press at the end of March, April and May, preceding the maximum Negro river stage, which usually occurs between June and July.

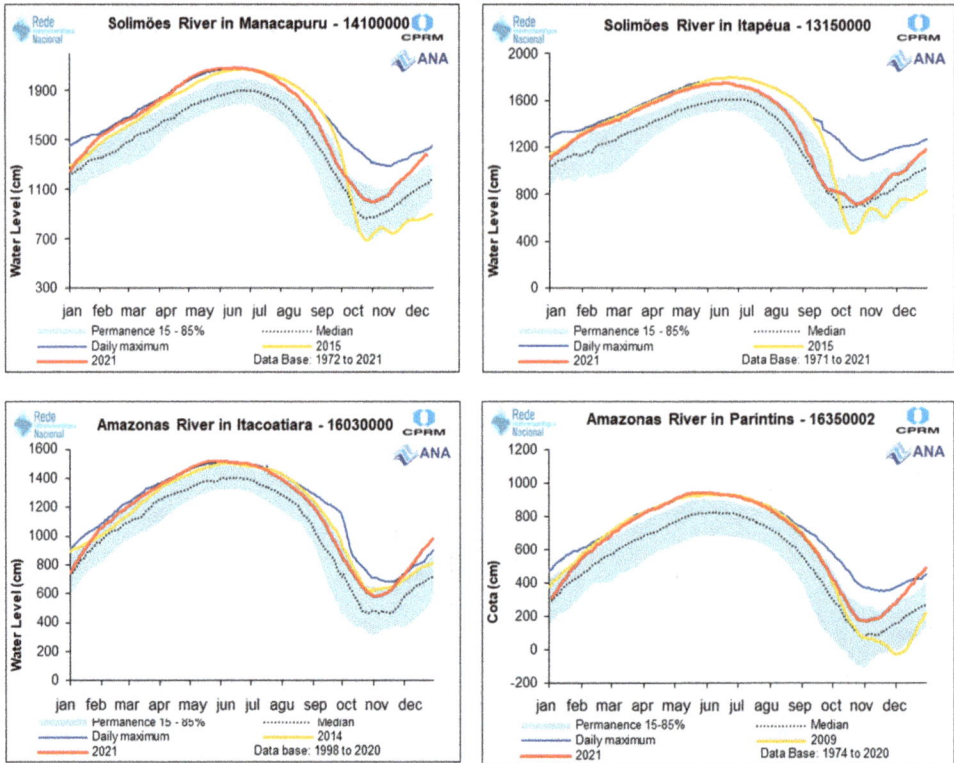

Figure 5:   Manacapuru, Itapéua, Itacoatiara and Parintins graphs with a similar behavior in 2021. The yellow lines represents the year that the maximum level was observed before 2021. See Fig. 4 for Manaus graph.

The flood alert system in Manaus is a non-structural measure adopted in order to minimize the damage caused by floods and leaks in river watershed. The operation of the flood alert system in Manaus has been developing positively, with institutional cooperation from the Civil Defense, the State Government, the City Hall and the various press vehicles.

An average flood year, Negro river has increased flow in the months of May and June. Thus far Solimões has this increase in March and April. Already in an above-average flood year, the flow of Solimões grows in February.

Since 2007, the Amazonas Hydrological Alert System has been publishing the "Western Amazon Hydrometeorological Monitoring Bulletin". The bulletin provides updated information on fluviometric stations considered strategic that make up the National Hydrometeorological Network, under the responsibility of the National Water and Basic Sanitation Agency (ANA), operated by the Geological Survey of Brazil (SGB-CPRM).The rivers levels are presented, in comparison to the data of the respective historical series, in the form of maps, graphs, figures and text, in order to facilitate the understanding by the user public, composed mainly by the local press, government agencies, universities and the general population.

In addition to monitoring the level of the rivers, a climatological monitoring of the last 30 days is also presented, accompanied by a forecast of precipitation for the next 15 days, in which the data are discretized according to the large basins of the Amazon. The rainfall monitoring data are obtained, organized and interpreted by meteorologist from the Amazon Protection System (Manaus Regional Center), through satellite data.

In 2021, many of the stations monitored in the bulletin reached historic records, which were communicated through bulletins. Table 2 presents the maximum levels observed in 2021, as well as the largest floods observed previously. Highlighted in red are the stations in which the previous records were surpassed in 2021. They are: São Gabriel da Cachoeira, Barcelos and Manaus (Negro river), Manacapuru (Solimões river), Careiro da Várzea (Amazon river basin), Itacoatiara and Parintins (Amazon river), all located in the state of Amazonas.

Table 2:   Previous historical maximum quotas and those reached in 2021. Highlights indicate the stations in which the previous records were surpassed in 2021.

Station (river)	Year related to the maximum quote		Year 2021		
	Date	Quote (cm)	Date of the maximum level	Maximum level (cm)	Comparison to the maximum level before (cm)
Barcelos (Negro)	**13/06/76**	**1,032**	**27/06/21**	**1,046**	**14**
Beruri (Purus)	24/06/15	2,236	29/06/21	2,198	−38
Boa Vista (Branco)	08/06/11	1,028	10/06/21	856	−172
Caracaraí (Branco)	09/06/11	1,114	12/06/21	947	−167
Careiro (P. Careiro)	**30/05/12**	**1,743**	**06/06/21**	**1,746**	**3**
Fonte Boa (Solimões)	06/06/15	2,282	21/05/21	2,218	−64
Humaitá (Madeira)	11/04/14	2,563	10/04/21	2,248	−315
Itacoatiara (Amazonas)	**19/06/09**	**1,505**	**28/05/21**	**1,520**	**15**
Itapeuá (Solimões)	24/06/15	1,801	21/06/21	1,750	−51
Manacapuru (Solimões)	**25/06/15**	**2,078**	**17/06/21**	**2,086**	**8**
Manaus (Negro)	**29/05/12**	**2,997**	**16/06/21**	**3,002**	**5**
Parintins (Amazonas)	**31/05/09**	**936**	**21/05/21**	**946**	**10**
Rio Branco (Acre)	05/03/15	1,834	17/02/21	1,578	−256
São Gabriel da Cachoeira (Negro)	**20/07/02**	**1,217**	**11/06/21**	**1,268**	**51**
Tabatinga (Solimões)	28/05/99	1,382	07/05/21	1,282	−100
Santa Isabel do Rio Negro, Tapuruquara	02/06/76	890	17/06/21	843	−47

Geological Survey of Brazil forecasts are based on a simple linear regression, where the maximum annual stages are correlated with the forecast dates, which are 31 March, 30 April and 31 May [6]. In 2021, since the first flood alert, forecasts have indicated the possibility of a significant flood event in the studied stations. The second and third forecasts indicate an increasing probability of one of the biggest floods in the hydrological monitoring history in some cities of Amazonas. The observed results confirmed the forecasts presented, marking the year 2021 as the record flood (Fig. 6).

1st Forecast 2021	31 March	29.45 m (maximum 30.35 m)
2nd Forecast 2021	30 April	30.00 m (+/− 50 cm)
3rd Forecast 2021	31 May	30.00 m (+/− 2 cm)

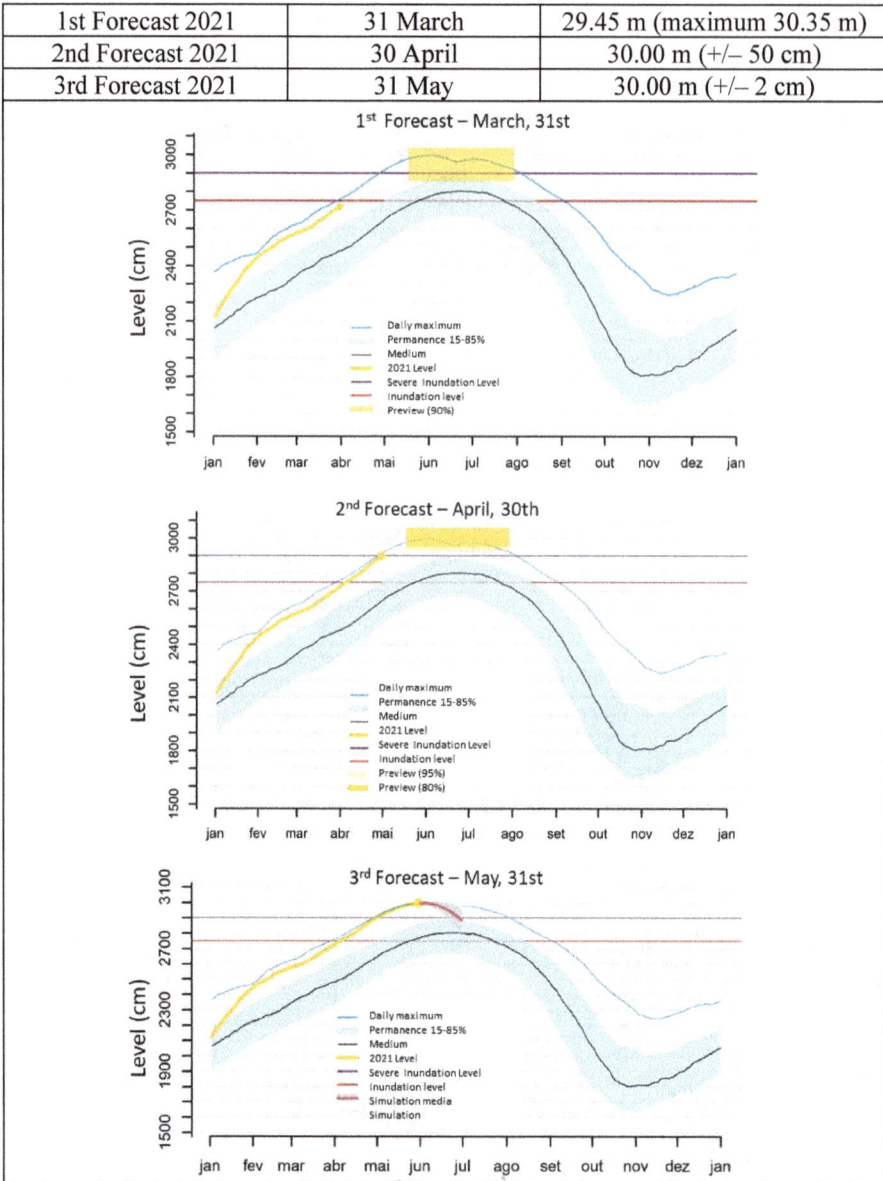

Figure 6: The flood forecast presented by Geological Survey of Brazil in 2021.

5 CONCLUSIONS

To analyze the flooding at the Manaus Station, it is necessary to monitor the evolution of the level of the rivers rising in the different stations, to know the difference between two important river flows (Negro and Solimões), in addition to understanding the dynamics of the entire basin and the other events flooding occurred on the Negro River at the Manaus Station. Considering the important historical series recorded at this station and the use of static analysis methods with hydrological data, it was possible perform a flood forecasting model for the Manaus Station on the Negro river.

In 2021 some monitored stations in the Amazon reached river levels higher than the maximum observed until then, making this year the biggest flood in the entire history of monitoring in the state of Amazonas. This event had the contribution of a La Niña event, which increased precipitation in different regions of Amazon Basin.

Geological Survey of Brazil forecasts have indicated the possibility of a significant flood event in the studied stations. The observed results confirmed the forecasts presented (second and third), marking the year 2021 as the record flood.

In 2021, from the establishment of La Niña in the Western Amazon, increased the precipitation observed the north of the region during the rainy season from November 2020 to May 2021, as a result, we had the great flood event observed in 2021. Weather issues are considered in the forecasts, as they are reflected in the river level and in the time to the flood peak. By the way, in 2022 the Civil Defenses in Amazonas are following weather and hydrological forecasts in advance for the year flood alert.

ACKNOWLEDGMENT

The authors are grateful to all the hydrological team of the Geological Survey of Brazil in Manaus.

REFERENCES

[1]   Companhia de Pesquisa de Recursos Minerais/Serviço Geológico do Brasil (CPRM/SGB), Relatório de Cheia de 2012. Manaus, 2012. www.cprm.gov.br.

[2]   Trigg, M.A., Wilson, M.D., Bates, P.D., Horritt, M.S., Alsdorf, D.E., Forsberg, B.R. & Vega, M.C., Amazon flood wave hydraulics. *J. Hydrol.*, **374**, pp. 92–105, 2009. DOI: 10.1016/j.jhydrol.2009.06.004.

[3]   Getirana, A.C.V., & Paiva, R.C.D., Mapping large-scale river flow hydraulics in the Amazon Basin. *Water Resour. Res.*, **49**, pp. 2437–2445, 2013. DOI: 10.1002/wrcr.20212.

[4]   Meade, R.H. et al., Backwater effects in the Amazon River basin of Brazil. *Environ. Geol. Water Sci.*, **18**, pp. 105–114, 1991. DOI: 10.1007/BF01704664.

[5]   Filizola, N. & Guyot, J.L., Suspended sediment yields in the Amazon basin: An assessment using the Brazilian national data set. *Hydrological Processes*, **23**(22), pp. 3207–3215, 2009.

[6]   Maciel, J.S.C., Simões, L.G.A., Corrêa, B.G.S., Carvalho, I.M.R. & Oliveira, M.A., Flood forecast in Manaus, Amazonas, Brazil. *FRIAR 2020*, vol. 194, WIT Press: Southampton and Boston, p. 63, 2020.

[7]   Vale, R., Filizola, N., Souza, R. & Schongart, J., A cheia de 2009 na Amazônia Brasileira. *Revista Brasileira de Geociências*, **4**. Pp. 577–586, 2011.

[8]   Barichivich, J., Gloor, E., Peylin, P., Brienen, R.J.W., Schöngart, J., Espinoza, J.C. & Pattnayak, K., Recent intensification of Amazon flooding extremes driven by strengthened Walker circulation. *Science Advances*, 2018.

[9]   Chevuturi, A., Klingaman, N.P., Rudorff, C.M., Coelho, C.A.S. & Schöngart, J., Forecasting annual maximum water level for the Negro River at Manaus. *Climate Resilient and Sustainability*, pp. 1–17, 2021. DOI: 10.1002/cli2.18.

[10] Boers, N., Bookhagen, B., Barbosa, H., Marwan, N., Kurths, J. & Marengo, J.A., Prediction of extreme floods in the eastern Central Andes based on a complex networks approach. *Nat Commun.*, **5**, p. 5199, 2014. DOI: 10.1038/ncomms6199.

[11] Espinoza, J.C., Ronchail, J., Frappart, F., Lavado, W., Santini, W. & Guyot, J.L., The major floods in the Amazonas River and tributaries (Western Amazon basin) during the 1970–2012 period: A focus on the 2012 flood. *Journal of Hydrometeorology*, **14**, pp. 1000–1008, 2013. DOI: 10.1175/JHM-D-12-0100.1.

[12] Espinoza, J.C., Marengo, J.A., Schongart, J. & Jimenez, J.C., The new historical flood of 2021 in the Amazon River compared to major floods of the 21st century: Atmospheric features in the context of the intensification of floods. *Weather and Climate Extremes*, **35**(11), 100406, 2021. DOI: 10.1016/j.wace.2021.100406.

[13] Sena, J.A., Beser de Deus, L.A., Freitas, M.A.V. & Costa, L., Extreme events of droughts and floods in Amazonia: 2005 and 2009. *Water Resources Management*, **25**, pp. 110–125, 2012.

INCORPORATION OF FLOW VELOCITY IN FLOOD DAMAGE ESTIMATION: AHR RIVER VALLEY 2021 STUDY, GERMANY

EVA PAŽOURKOVÁ, MARTIN SALAJ & RICARDO E. WONG MONTOYA
Aon Impact Forecasting, Czech Republic

ABSTRACT

A common practice when catastrophe model vulnerability component is being developed is the use of inundation depth as a hazard metric. However, other quantities associated with inundated waters could also yield a significant effect on estimated flood loss being widely published. The key aspect of the study is to assess the relevance of the water depth, flow velocity, and their combinations in the damage estimates done by catastrophe models. Could the flow velocity in the loss calculation process provide some benefits to the catastrophe loss modelling process? In July 2021 the western part of Germany was affected by serious flooding. The physical conditions of the inundated buildings were evaluated in the frame of the Copernicus EMS satellite-based damage assessment. The two heavily impacted areas, the towns of Schuld and Altenahr, were selected for further re-simulation aiming to test the effect of depth and velocity. In the first step, horizontal velocity and flood depth were calculated for each building using a 2D hydraulic simulation. Subsequently, structural damage models were analysed with focus on their predictive skills and variability. A force-based threshold was selected to calibrate the total loss probability, as a feature of the new vulnerability component. As a result, a hazard intensity metric expressed as combined parameter of water depth and flow velocity is finally implemented in the catastrophe model. When comparing the proposed solution with traditional depth-based approach, one can see a slight increment in the modelled monetary damage and a significantly better correlation with the observed damage identified in the study area. The embedded effect of velocity could therefore improve the accuracy and sensitivity of catastrophe flood models, particularly in high-slope areas and in events with extreme and short rainfall intensities to sudden increments in the building damage level, as assessed for the flood in the Ahr valley.

Keywords: flow velocity, flood depth, catastrophe flood modelling, structural stability.

1 INTRODUCTION

The flood damage estimation is performed by the vulnerability component of catastrophe models, employing damage curves that assign a damage ratio for a given hazard intensity. This ratio corresponds to the fraction of repair cost or monetary loss with respect to the sum insured of the asset. For the definition of the hazard parameter, it is a common practice the usage of inundation depth (Smith [1]). Nevertheless, the depth solely might not reflect the damaging processes occurring in the different flood types (Paprotny et al. [2]), such as flash floods where the flow velocity is also an important parameter in the damaging process.

The correlation between flow velocity and building damage could not be analysed independently from the inundation depth. Previous studies have recognized the energy head as a suitable parameter for the forecast of structural damage of buildings (Schwarz and Maiwald [3], Kreibich et al. [4]). Additionally, the flow velocity could influence the loss calculation process in certain contexts, such as levee-break floods or mountainous flood plains (Gallegos et al. [5]).

Several structural damage models have assessed the compound effect of inundation depth and flow velocity over different building types (Black [6], Dale et al. [7]). The models collated in Gallegos et al. [5] and Smith et al. [8] consider a threshold-type criterion for the definition of the collapse condition, being ruled by force-, energy head- or discharge-

WIT Transactions on The Built Environment, Vol 208, © 2022 WIT Press
www.witpress.com, ISSN 1743-3509 (on-line)
doi:10.2495/FRIAR220021

thresholds. In this sense, for the scenarios where the hazard intensity exceeds the threshold established by the model, the collapse or total loss of the building is expected (Smith et al. [8]).

The objective of this paper is to evaluate the influence of flow velocity in the loss modelling process, by assessing a hazard intensity metric that combines the effect of inundation depth and flow velocity and comparing it with the inundation depth-based damage curves.

2 STUDY AREA AND DATA

2.1 Study area

The Ahr river basin is situated in the Eifel region (low mountain range) in the western part of Germany. The river Ahr rises in the municipality Blankenheim at an elevation of approximately 470 m above sea level in the state of North Rhine-Westphalia. The length of the river is 85.1 km after roughly 18 km from the source it crosses the borders of federal states from North Rhine-Westphalia into Rhineland-Palatinate. The Ahr and its tributaries are the main drainage system of the eastern Eifel with a size of the catchment of about 896.4 km^2. The Ahr river, as a left tributary, flows into the Rhein at an elevation approximately 50 m above sea level (Roggenkam and Herget [9]). The Eifel region has relatively shallow soils developed on the paleozoic siliciclastic sedimentary rocks. The topography is characterized by plain surmounted by individual mountain ranges. The river network is deeply cut into valleys which usually have a V-shape, steep slope and they are narrow which could cause funnel-like effects in the flood events. The geological, pedological, and topographical conditions of the area could cause quite a fast and intensive hydrological response (Kreienkamp et al. [10]).

In July 2021, Western and Central Europe was affected by torrential rainfall caused by a slow-moving area of low-pressure named Bernd. A combination of events related to low pressure Bernd, high moisture in the air, which was constantly tapped from the Mediterranean Sea, and stop of movement almost for four consecutive days, led to extreme rainfall. Many areas of western Germany recorded rainfall rates that exceeded a 1-in-100-year return period in some areas even 1-in-1000-year return period. Unfortunately, the soil was already almost saturated from previous rainfall events and a combination of all factors caused catastrophic floods on the Ahr river but also on the other parts of Germany and Europe. This event is ranked among the deadliest natural disasters in modern German history and caused a huge financial loss. The towns Schuld and Altenburg, which are located on the Ahr river and were heavily affected, were selected for a detailed analysis (Cat Alert [11]).

2.2 Copernicus EMS grading assessment

Copernicus is an EU programme aimed to provide information services based on satellite Earth observation and in situ data. The Copernicus Emergency Management Service (CEMS) provide mapping service in cases of natural disasters, human-made emergencies, and humanitarian crises during all phases of the emergency management cycle (forecast, notifies, monitor) throughout the world. The satellite imagery and other geospatial data are used for developing the products. The rapid mapping activations could have four kind of products: reference map, first estimate map, delineation map, and grading map. This geospatial information or products are provided within hours or a few days after disaster (Dorati et al. [12], CEMS [13]).

The grading map is based on post-event imagery and provides an assessment of the damage grade and its spatial distribution. Unfortunately, it is not possible to use the existing methodologies for damage assessment of building structures because most of them are designed for ground-based field evaluations. The CEMS for rapid damage assessment of building structures involved, as a starting point, the European Macroseismic Scale 1998 (EMS-98). Due to challenges and limitations related to geospatially based damage assessment as the bird's-eye view, the image geometric resolution, the radiometric resolution, and the subjectivity of the interpretation the CEMS adapted the EMS-98 classes. The categories were aggregated and simplified to reflect the inherent limitations of the technology adopted (remotely sensed imagery) and time constraints in place (rapid, near real-time). Two additional categories (possibly damage, no visible damage) are added to consider the intrinsic uncertainty of the methodology. The CEMS products are made from remotely sensed imagery and in almost real-time, not as ground observation data therefore it is necessary to take into account a certain uncertainty in their use (Dorati et al. [12], CEMS [13]).

Table 1:   Comparison of Copernicus EMS and EMS-98 classes for damage assessment of buildings [12].

Copernicus EMS classes		EMS-98 classes
No visible damage		Grade 0: No damage
Possibly damaged		It refers to cases when the confidence level of the interpretation is slightly lower (e.g. bad image quality)
Damaged		Grade 1: Negligible to slight damage
		Grade 2: Moderate damage
		Grade 3: Substantial to heavy damage
Destroyed		Grade 4: Very heavy damage
		Grade 5: Destruction

The EMSR517 flood in Western Germany is a product of the CEMS Rapid Mapping Service to monitor the flood evolution after heavy rain which affected federal state Rhineland-Palatinate. This product consists of 11 delineation products (43 maps) and 16 grading products (37 maps), particularly the [EMSR517] Bad Neuenahr-Ahrweiler: Grading Product, Monitoring 1, version 3, release 1, RTP Map#01 published on 11 August 2021 was used in this study. The number of buildings classified according to CEMS classes is 753 destroyed, 5078 damaged and 1175 possibly damage in this area (CEMS [13]).

2.3 2D hydraulic simulation

Hydraulic simulation of the July 2021 flood was done as 2D model with 10m resolution by HPC TUFLOW software from BMT [14]. TUFLOW is a computer program for simulating depth-averaged, one and two-dimensional free-surface flows such. Flow regimes are handled by adaptation of the Saint Venant Equations and Shallow Water Equations with momentum and continuity equations for free flow.

Source of elevations was "DGM10 – Digitales Geländemodell Gitterweite 10 m". Shape of buildings were used from the OSM [15]. The simulation considered 50% blockage of bridges. Flows and water surface elevations were considered based on analysis of hydrology data from webpages of Landesamt für Umwelt (LfU) [16] and Bundesanstalt für Gewässerkunde (BfG) [17], which were publicly available in August 2021 (limited number of records; without detailed verification of providers). The record of the gauging station Müsch located in the Ahr river valley is shown on Fig. 1 as an example.

Figure 1:   Gauging station Müsch on Ahr river – Discharges during flood event 14–15 July 2021 and official discharges for flood of return period 20, 50 and 100 years. *(Source: Values from Landesamt für Umwelt (LfU) [16].)*

The values of water depth–flow velocity for each one of the building's locations are obtained from the hydraulic simulation. In Fig. 2, each point of the scatter represents one of the buildings mapped by Copernicus EMS. Additionally, the plot shows the damage grade (possibly damaged, damaged, destroyed) assigned to each building of the dataset.

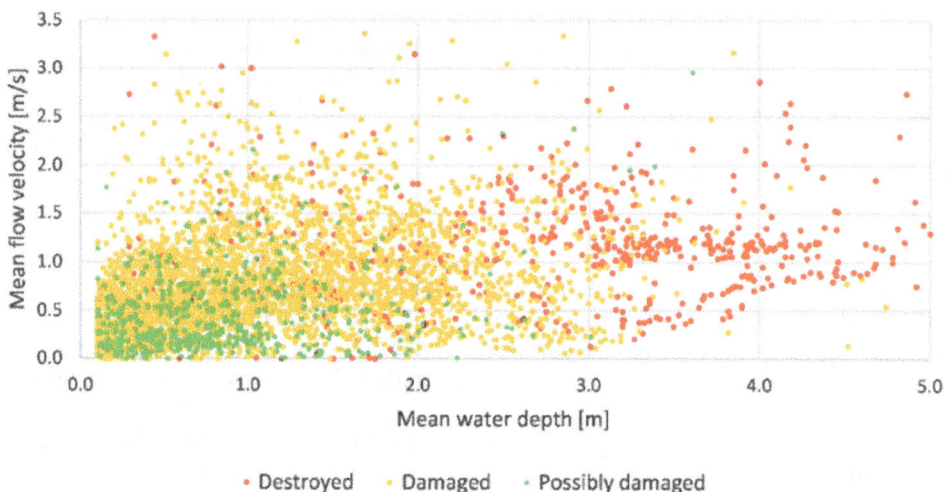

Figure 2:   Water depth and flow velocity calculated for each building in the area of the CEMS activation and their corresponding damage grade.

3 METHODS

3.1 Structural stability of buildings

The stability of buildings could be threatened due to several mechanisms, including lateral pressures from depth differentials and flood velocities, buoyancy, debris impact, and scour. These aspects have been analysed in the development of different damage models proposed in the scientific literature (Gallegos et al. [5]). In Table 2, different damage threshold-type criteria are shown. The models are defined based on either water depth h, flow velocity V, discharge per unit width q, energy head E, and force F (per unit width and fluid density).

- Discharge per unit width

$$q = hV, \tag{1}$$

- Energy head

$$E = h + V^2/2g, \tag{2}$$

- Indicator of flow force

$$F = hV^2. \tag{3}$$

Table 2: Review of existing structural damage models.

Failure criteria	Country	Threshold definition	Construction type	Ref.
WRL Light Str.	Australia	$q > 1\,m^2/s, h > 2m, V > 2\,m/s$	–	[8]
WRL Max.		$q > 4\,m^2/s, h > 4m, V > 4\,m/s$		
Black, 1 storey	USA	Fit of exponential decay function [5]	Wood frame	[6], [5]
Black, 2 storeys				
Clausen and Clark	UK	$q > 7\,m^2/s, V > 2\,m/s$	–	[5]
Kreibich, et al., Energy-based	Germany	$E > 2m$	–	[4], [5]
Kreibich, et al., Depth-based		$h > 2m$		
Kreibich, et al., Discharge-based		$q > 1.5\,m^2/s$		
Kreibich, et al., Force-based		$F > 2\,m^3/s^2$		
CH2M, 1 storey	USA	$F > 7.6\,m^3/s^2, h > 3m$	Wood frame	[5]
CH2M, 2 storeys		$F > 7.6\,m^3/s^2, h > 4.5m$		
Gallegos et al., Moderate	USA	$F > 0.75\,m^3/s^2$	Wood framed	[5]
QRA	Australia	$q > 1.2\,m^2/s, h > 2m, V > 2\,m/s$	–	[8], [18]
NCC Light Str.	Australia	$q > 1\,m^2/s, h > 2m, V > 2\,m/s$	Masonry, wood	[8], [19]
NCC Heavy Str.		$q > 2.5\,m^2/s, h > 2.5m, V > 2.5\,m/s$	Steel frame, concrete	

The structural damage models collated in Table 2 are plotted in Fig. 3 to allow the comparison of the curves. The figure highlights the overall uncertainty in the matter of building stability under flooding conditions (Smith et al. [8]). The difference between the models is due to the characteristics of the buildings being assessed in each study. The vulnerability of the buildings is linked to their construction type, which play an important role in the definition of the threshold from which the building is prone to collapse.

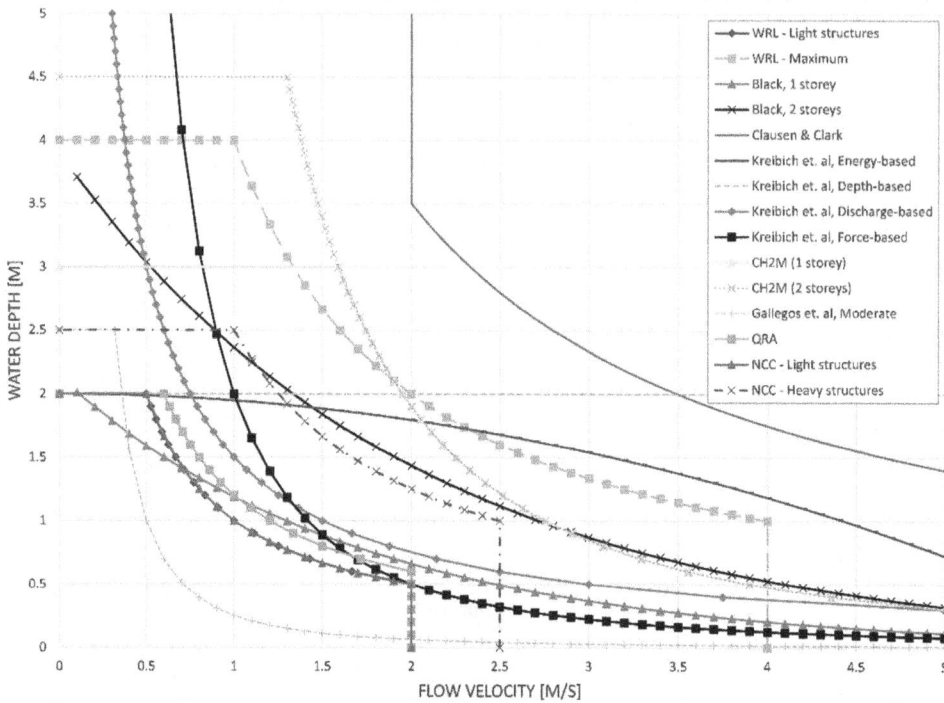

Figure 3: Comparison of building stability curves considered for this study. The failure condition is expected for the depth–velocity combination above the curves. *(Source: Adapted from Gallegos et al. [5].)*

To evaluate the performance of the models, regarding their capability to predict the failure of the buildings in the study area, the number of hits H, misses M, false alarms FA, and correct rejections Z are quantified. The definition of each one of these parameters is shown in Table 3. The failure condition of the building is taken from the class Destroyed from the Copernicus EMS damage assessment. The model outcome corresponds to the predicted failure or non-failure of the buildings analysed with each stability curve (Gallegos et al. [5], Baldwin and Kain [20]).

Table 3: Contingency table for assessment of predictive skill of building stability curves.

	Observed condition – Copernicus EMS grading	
Forecast – Model outcome	Failure	No failure
Failure predicted	Hit (H)	False alarm (FA)
Failure non predicted	Miss (M)	Correct rejection (Z)

Furthermore, the following metrics were computed (eqns (4)–(8)):

- Probability of detection (POD)

$$POD = H/(H + M), \tag{4}$$

- Probability of false detection (POFD)

$$PFOD = FA/(Z + FA), \tag{5}$$

- False alarm ratio (FAR)

$$FAR = FA/(H + FA), \tag{6}$$

- Bias (B)

$$B = (H + FA)/(H + M), \tag{7}$$

- Skill (S)

$$S = POD - POFD. \tag{8}$$

3.2 Correlation analysis and hazard intensity metric

The analysis of the dependencies between the hazard metrics and their dependency with the structural damage is performed for the dataset of impact parameters on the buildings in the study area (Fig. 2).

As an outcome of the failure condition assessment and the analysis of the correlation between the hazard parameters and the structural damage, different alternatives for the hazard intensity metric as a function of water depth and flow velocity are tested. The output regarding the effect on the monetary losses is then evaluated by comparison with the losses obtained with the depth-based approach.

4 RESULTS AND DISCUSSION

4.1 Performance of stability models

The capacity to forecast the structural failure of each model is reflected in the damage metrics shown in Table 4. Several models reach a skill value of more than 0.40, being the highest skill value ($S = 0.48$) obtained with the discharge-based models WRL – Light structures and QRA. Nevertheless, both models are within the ones with the greatest bias. The force- and energy-based models proposed by Kreibich et al. [4] reach similar metrics of skill, FAR and bias. Additionally, discharge-based model NCC – Heavy structures performed analogously. Finally, we could not conclude that even energy, discharge, or force models account for a better performance than the others.

The skill values obtained are low in comparison with previous studies, such as Gallegos et al. [5]. This could be related to many uncertainty factors, including the damage assessment performed by photo interpretation, that does not reach the same output as an evaluation in the field (ground truth). Furthermore, the category destroyed of the CEMS grading does not correspond uniquely with washout buildings, but also considers the buildings in grade 4 of the EMS-98 classes, as shown in Table 1. Additionally, it is expected to have some discrepancies regarding the hydraulic simulation with respect to the real event.

The broad range of skill values obtained follows the high variability in the damage models shown in Fig. 3. As an example, Fig. 4 shows the zones of stability failure predicted by three of the models for the Schuld municipality. The model WRL for light structures proposed by Smith et al. [8] gives the greatest damage zone, covering most of the buildings identified as

Table 4:   Predictive skill, false alarm ratio and bias for each damage model applied in the Ahr river.

	Predictive skill (S)	False alarm ratio (FAR)	Bias (B)
WRL – Light structures	0.48	0.89	7.56
WRL – Maximum	0.26	0.62	0.75
Black, 1 storey	0.46	0.89	7.21
Black, 2 storeys	0.39	0.77	2.06
Clausen and Clark	0.06	0.52	0.13
Kreibich et al., Energy-based	0.42	0.83	3.44
Kreibich et al., Depth-based	0.40	0.82	3.11
Kreibich et al., Discharge-based	0.45	0.86	4.65
Kreibich et al., Force-based	0.42	0.84	3.74
CH2M (1 storey)	0.31	0.65	0.98
CH2M (2 storeys)	0.22	0.61	0.62
Gallegos et al., Moderate	0.42	0.89	7.20
QRA	0.48	0.88	6.65
NCC – Light structures	0.48	0.89	7.56
NCC – Heavy structures	0.47	0.79	2.88

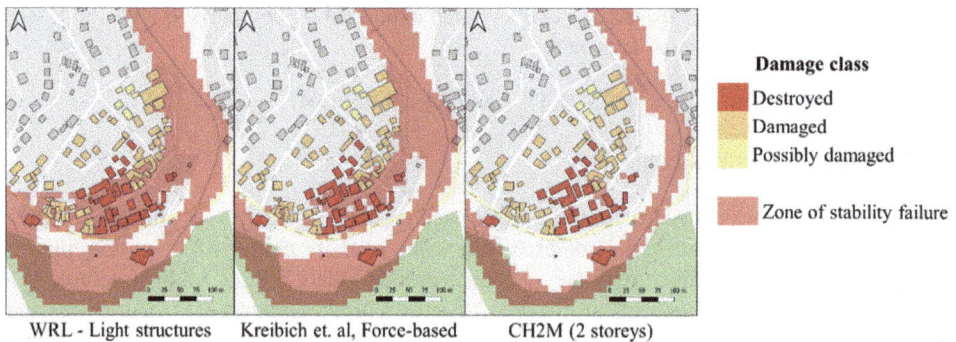

WRL - Light structures   Kreibich et. al, Force-based   CH2M (2 storeys)

Figure 4: Zones of stability failure predicted by the damage models for the Schuld area.

destroyed by the Copernicus grading assessment. Nevertheless, some of the damaged buildings are also included in this zone. Furthermore, a smaller area is obtained with the model proposed by Kreibich et al. [4], while the CH2M model is not capable to map the area of the destroyed buildings, underestimating the damage zone.

A unitary force-based model is selected to be implemented in the model. The calibration of the optimal force threshold is performed based on the expected chance of total loss (COTL). In this sense, the flow velocity influences the estimation of the proportion of buildings that could reach a failure condition. The COTL is further applied in the damage function, corresponding to the condition of mean damage ratio equal to 1.

4.2 Correlation and predictive skill analysis

The water depth has a significant effect on the losses. The depth above approximately 2 m has a big chance of destroying the buildings (see Figs 2 and 5). The threshold when the flow velocity has an impact on the building structural damage could be approximately 1 m/s but there is no clear limit for the flow velocity between damaged and destroyed buildings.

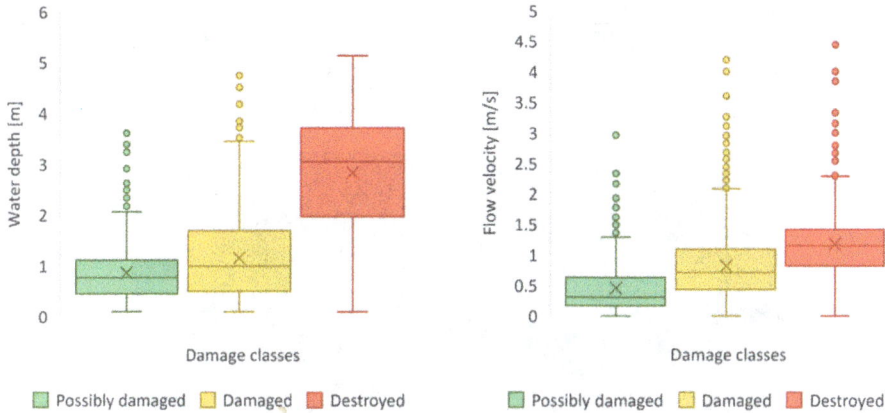

Figure 5: Boxplots for the calculated water depth and flow velocity at the building's location, categorized by damage classes.

As shown in Table 5, the flow velocity has the strongest correlation relationship with the indicator flow force and the intensity. The water depth has the strongest correlation with the energy head, as the potential energy is the dominant parameter for the energy head.

Table 5: Correlation between the impact parameters.

	Flow velocity (m/s)	Water depth (m)	Energy head (m)	Indicator flow force (m³/s)	Intensity (m²/s)
Flow velocity (m/s)	1.00	0.35	0.40	0.92	0.82
Water depth (m)		1.00	0.99	0.66	0.80
Energy head (m)			1.00	0.70	0.83
Indicator for flow force (m³/s)				1.00	0.97
Intensity (m²/s)					1.00

For each hazard parameters, minimum thresholds for experiencing certain level of damage were established. The skill of each one of these impact parameters to identify the damage class of the buildings in the municipalities of Altenahr and Schuld is shown in Table 6. The flow velocity has the lowest skill value. In contrast, energy head, which considers the effect of the water depth and flow velocity (potential and the kinetic energy, respectively), reached the highest predictive capability. It is evidenced that, for these specific locations, the flow velocity as part of the energy head has impact on the damage of the buildings.

Table 6: Predictive skill, false alarm ratio and bias for each alternative of hazard intensity applied in the municipalities of Altenahr and Schuld.

Location		Flow velocity (m/s)	Water depth (m)	Energy head (m)	Indicator flow force (m³/s)	Intensity (m²/s)
Altenahr (n = 49)	Skill (S)	0.03	0.24	0.43	0.16	0.32
	FAR	0.33	0.29	0.21	0.29	0.26
	Bias (B)	0.75	1.41	1.06	0.97	1.22
Schuld (n = 41)	Skill (S)	0.07	0.21	0.26	0.16	0.21
	FAR	0.47	0.33	0.35	0.36	0.33
	Bias (B)	0.85	0.60	0.85	0.55	0.60

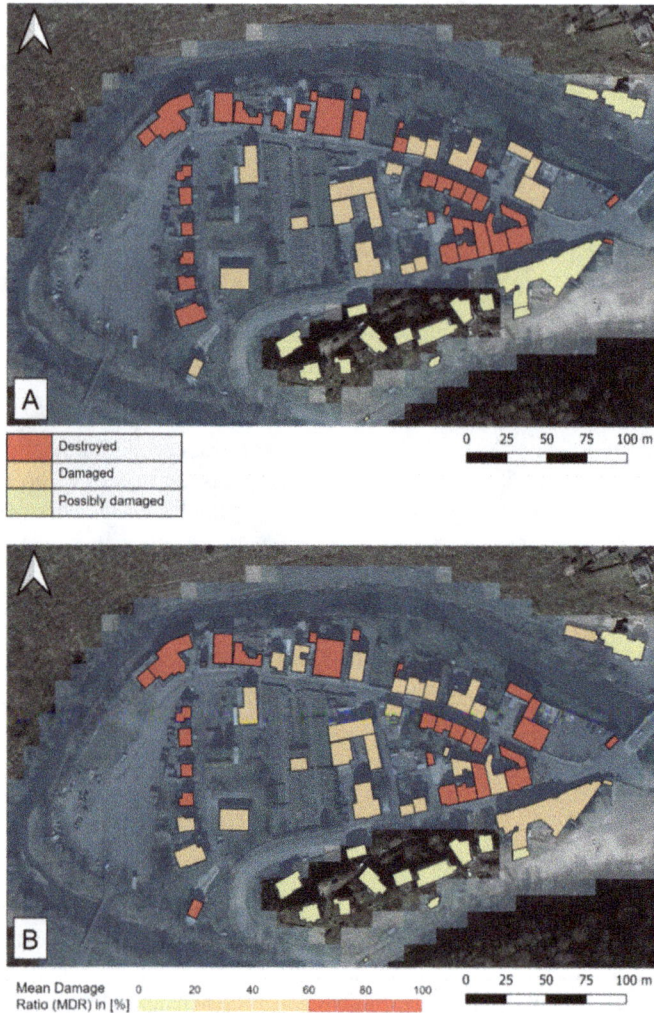

Figure 6: Location: Altenahr municipality. (A) Copernicus EMS damage assessment; and (B) Application of the hazard intensity metric as a function of water depth and flow velocity. MDR = f (depth, velocity).

Further research of different types of floods and their impact on the structural damage of buildings is necessary. The methodology applied (for the determination of the damage grade or in the hydraulic simulation) or the influence of the impact parameters not considered in this study (impact of debris or flood duration) represent sources of uncertainty in the analysis.

4.3 Performance of the hazard intensity metric

The hazard metric is finally implemented in the catastrophe model. Due to the lack of information about the specific buildings, for this study the building stock remains undifferentiated, considering masonry buildings for the damage curve definition. For the Altenahr municipality, in Fig. 6 a partially good agreement is reached between the Copernicus EMS damage assessment and the distribution of expected monetary loss given by the mean damage ratio.

Furthermore, it was obtained an increment of 2%–4% on the expected monetary loss in comparison with the depth-based approach. The slight increment in the monetary loss implies a small impact in the probable maximum loss (PML) calculated by the model. Nevertheless, the impact is localized in small part of the floodplain, particularly high slope areas, being the effect on the average annual loss (AAL) significant for these specific locations and allowing a better estimation in the underwriting process.

5 CONCLUSIONS

The flow velocity incorporation in the flood damage estimation process is achieved based on the study of its influence on the stability of the buildings and the dependency of the structural damage and the hazard intensity parameters.

Regarding the existing structural damage models, the highest skill in the prediction of the destroyed buildings was obtained by the WRL – Light structures model, while the model proposed by Clausen and Clark has the worst skill value. Nevertheless, it was recognized that in most of the cases, high skill models implied high bias, with an overestimation in the number of destroyed buildings.

The results from the analysis performed in this study showed that the energy head is a suitable parameter for the forecasting of structural damage, as suggested by Kreibich et al. [4]. Further study in this matter is necessary, including also different flood types and areas with different morphological conditions.

The hazard intensity metric defined based on water depth and flow velocity implemented in the catastrophe model derived in a slight increment in the monetary losses compared to the depth-based approach. This difference could increase in the case of high-slope areas.

The vulnerability component developed in this paper aimed to consider the combined effects of water depth and flow velocity on the monetary damage that a building can undergo. Furthermore, the construction type of a building should be considered since it is related to its vulnerability and probability of collapse.

There are many uncertainty factors faced in this study, including limitations of the Copernicus EMS product and the hydraulic simulation to reflect the real event conditions. Finally, the current paper is part of an ongoing study. An evaluation of the transferability of the model and calibration of the monetary losses would generate further improvements in the modelling process.

REFERENCES

[1]   Smith, D.I., Flood damage estimation: A review of urban stage damage curves and loss functions. *Water SA*, **20**(3), pp. 231–238, 1994.

[2]   Paprotny, D., Kreibich, H., Morales-Nápoles, O., Wagenaar, D., Castellarin, A., Carisi, F., Bertin, X., Merz, B. & Schröter, K., A probabilistic approach to estimating residential losses from different flood types. *Natural Hazards*, **105**, pp. 2569–2601, 2021.

[3]   Schwarz, J. & Maiwald, H., Damage and loss prediction model based on the vulnerability of building types. *4th International Symposium on Flood Defence*, 2008.

[4]   Kreibich, H., Piroth, K., Seifert, I., Maiwald, H., Kunert, U., Schwarz, J., Merz, B. & Thieken, A.H., Is flow velocity a significant parameter in flood damage modelling? *Nat. Hazards Earth Syst. Sci.*, **9**(5), pp. 1679–1692, 2009.

[5]   Gallegos, H., Schubert, J. & Sanders, B., Structural damage prediction in a high-velocity urban dam-break flood: Field-scale assessment of predictive skill. *Journal of Engineering Mechanics*, **138**(10), pp. 1249–1262, 2012.

[6]   Black, R., Flood proofing rural residences. A Project Agnes report, Cornell University: New York, 1975.

[7]   Dale, K., Edwards, M., Middelmann, M. & Zoppou, C., Structural flood vulnerability and the Australianisation of Black's curves. *Proceedings of the Risk Conference*, 2004.

[8]   Smith, G.P., Davey, E.K. & Cox, R.J., Flood hazard. WRL Technical Report 2014/07, UNSW: New South Wales, 2014.

[9]   Roggenkamp, T. & Herget, J., Reconstructing peak discharges of historic floods of the river Ahr, Germany. *Erdkunde*, **68**(1), pp. 49–59, 2014.

[10]   Kreienkamp, F., Philip, S.Y., Tradowsky, J.S., Kew, S.F., Lorenz, P., Arrighi, J. & Wanders, N., Rapid attribution of heavy rainfall events leading to the severe flooding in Western Europe during July 2021. *World Weather Attribution*, p. 51, 2021.

[11]   Cat Alert, Western and Central Europe Floods, Impact Forecasting, Aon, 2021.

[12]   Dorati C., Kucera, J., Marí i Rivero I. & Wania, A., Product User Manual of Copernicus EMS Rapid Mapping. JRC Technical Report JRC111889, Joint Research Centre, 2018.

[13]   Copernicus Emergency Management Service (CEMS). https://emergency.copernicus.eu/. Accessed on: 25 Feb. 2022.

[14]   BMT, TUFLOW Classic/HPC User Manual, Build 2018-03-AD. BMT: Australia, p. 723, 2018

[15]   OSM, https://www.openstreetmap.org/.

[16]   LfU, http://213.139.159.46/prj-wwvauskunft/projects/messstellen/wasserstand.

[17]   BfG, https://www.bafg.de/DE/06_Info_Service/01_Wasserstaende/wasserstaende_node.html.

[18]   Queensland Reconstruction Authority, Planning for stronger, more resilient floodplains. https://www.qra.qld.gov.au/publications-and-maps/building-and-planning-guides. Accessed on: 25 Feb. 2022.

[19]   Newcastle City Council, Newcastle flood policy 2003: A technical manual, 2003.

[20]   Baldwin, M. & Kain, J., Sensitivity of several performance measures to displacement error, bias, and event frequency. *Weather Forecast.*, **21**(4), pp. 636–648, 2006.

SIMULATIVE FLOOD DAMAGE MODELLING TAKING INTO ACCOUNT INUNDATION LEVEL AND FLOW VELOCITY: UNCERTAINTIES AND STRATEGIES FOR FURTHER REFINEMENT

HOLGER MAIWALD & JOCHEN SCHWARZ
Earthquake Damage Analysis Center (EDAC), Bauhaus Universität Weimar, Germany

ABSTRACT

In recent years, floods in Germany have caused billions of Euros in property damage. As part of the project "Innovative Vulnerability and Risk Assessment of Urban Areas against Flood Events" (INNOVARU), a realistic, practical model for the monetary assessment of potential flood damage to residential building stock was developed, which also allows the prognosis of structural damage. The structural damage can be predicted in the form of mean damage grades using vulnerability functions, which take into account the vulnerability of the different building types depending on the inundation level and flow velocity. So far, the scatter in the damage has not been taken into account. The paper presents "fragility functions" which enable the quantification of the exceedance probability of certain damage grades depending on inundation level and flow velocity. These functions allow the identification and implementation of the scatter of structural damage. They also enable a simulative damage prognosis using the Monte Carlo method, which provides the basis for loss calculations and serve to quantify the scatter within the financial loss indicators. This can introduce a new level of cost–benefit analyses for the planning of new flood protection measures. For lower flow velocities, typical for river floods, the study is based on a comprehensive qualified damage dataset compiled after the 2002 flood in Germany. The lack of reliable damage data caused by high flow velocities during flash flood events is compensated by an innovative approach. For this purpose, damage data from the tsunami of the Tohoku earthquake in Japan in 2011 are re-evaluated and included in the analysis. The developed "fragility functions" are applied to the re-interpretation of the August 2002 flood damage and loss in six different study areas in the Free State of Saxony. An outlook to the application for flash flood events is given.
Keywords: structural damage, damage prognosis, flow velocity, vulnerability classes, fragility functions, validation, simulation, scatter.

1 INTRODUCTION

The recent flood events in Germany (2002, 2010, 2013, 2021) show that even extreme events with very low probabilities of occurrence are possible and can result in devastating damage. The floods in 2002 and 2021 especially, have shown that in addition to pure penetration damage severe structural damage up to total destruction of the building substance could occur. Conventional flood loss models cannot consider this type of damage in an adequate way. An overview of such models is given in Jongman et al. [1].

Structural damage has mostly been examined in the past in terms of total failure of the structure depending on the influence of the flow velocity in connection with the water level [2], [3] or additional criteria for a partial failure are determined [4]. A refined differentiation to take into account the different damage patterns of structural damage, the conversion into concrete losses and the consideration of the scatter in the damage with comparable effects are missing in these studies.

The flood damage model developed at the Earthquake Damage Analysis Center (EDAC) of the Bauhaus-Universität Weimar provides with the derived vulnerability functions a first

WIT Transactions on The Built Environment, Vol 208, © 2022 WIT Press
www.witpress.com, ISSN 1743-3509 (on-line)
doi:10.2495/FRIAR220031

attempt to predict the structural damage in the form of mean damage grades depending on the building types, water level and specific energy height [5], [6].

The latest enhancements of the flood damage model refer to the insights from the project "Innovative Vulnerability and Risk Assessment of Urban Areas against Flood Events – INNOVARU", funded by the German Federal Ministry of Education and Research (BMBF). The aim of the project was the development of an application-ready model for the monetary assessment of the expected flood damage on the existing building stock and the decision and flood risk management support for the Free State of Saxony. The final approach is combining the advantages of the empirically supported prognosis of structural damage with the flexibility of building typological synthetic damage functions for loss estimations (cf. [7], [8]). As part of the sub-project processed by the EDAC, a set of refined functions for the prognosis of the structural damage is developed taking into account new combinations of water level, flow velocity and the building vulnerability, which are presented in Maiwald et al. [9], [10]. First results of the complete procedure developed in INNOVARU can be taken from Golz et al. [11].

Another result of the project are fragility functions with which the probability of exceedance for each damage grade can be determined. The new functions will be presented in the paper and applied to different study areas in Saxony on a microscale level for the 2002 flood.

2 BASIC ELEMENTS OF SIMULATIVE DAMAGE MODELING

2.1 Damage database

2.1.1 EDAC flood damage database
The EDAC flood damage database is the result of a comprehensive evaluation of damage caused by the 2002 flood in Saxony. Out of the 22,554 cases of flood-induced damage processed by the Saxonian Relief Bank (SAB) [12], detailed damage reports were available for around 8,000 buildings. Approximately 5,000 damage reports were accessed, analysed and entered into the database. In addition to the relevant building parameters and the inundation levels, the descriptions of the structural damage and the actual recovery costs were evaluated.

Based on hydraulic calculations (see also Section 2.3), estimated flow velocities were assigned to the individual damage cases (n ≈ 1,000) for four selected study areas in Saxony [13]. It has to be emphasized that the data in the EDAC flood damage database, for which flow velocities could be assigned, generally relate to damage cases with moderate flow velocities ($v_{max} \approx 2.5$ m/s) typical for river floods.

Reliable data for damage due to higher flow velocities (such as flash flood events) or the flow velocities for related damage cases contained in the database are currently not available. For this reason, an innovative approach is chosen to expand the damage database by including damage data from the 2011 Tohoku earthquake induced tsunami in Japan.

2.1.2 Tsunami damage data
A comprehensive damage database (n ≈ 252,000) is available for the tsunami after the 2011 "Tohoku earthquake" in Japan [14]. Based on this database, fragility functions for the tsunami hazard are derived in Suppasri et al. [15], [16] depending on the water level h on the building. Six damage classes were defined in analogy to the scheme of the European Macroseismic Scale EMS-98 [17] for earthquake shaking effects.

The damage classification proposed in Suppasri et al. [15], [16] was taken into account in Maiwald and Schwarz [18] to derive unified damage scales for the main natural hazard types. The database was re-evaluated in Maiwald and Schwarz [19] in order to develop a mathematically based methodology for deriving a vulnerability table of buildings (comparable to EMS-98) for tsunami impact. This leads to the assignment of vulnerability classes to the building types from the tsunami damage database. In addition, the flow velocities were estimated with respect to the coastal type (ria or plain coast).

2.2 Knowledge of building stock

For the INNOVARU project, the towns of Pirna, Grimma and Freital in the Free State of Saxony (in Germany) were selected as study areas. The parameters of the building stock of the towns of Döbeln, Eilenburg, Grimma and Flöha are available from previous research EDAC projects on a detailed microscale level. In the study areas, the entire building stock affected by the 2002 flood was systematically investigated on site; the relevant building parameters [20] and the existing flood marks were documented. The building stock of Pirna was analysed in Naumann and Rubin [7] and Naumann et al. [8], from one of the INNOVARU project partner. The building stock of the town Freital was documented within the INNOVARU project. The geographical location of the flood-affected areas considered in this paper can be taken from Fig. 1.

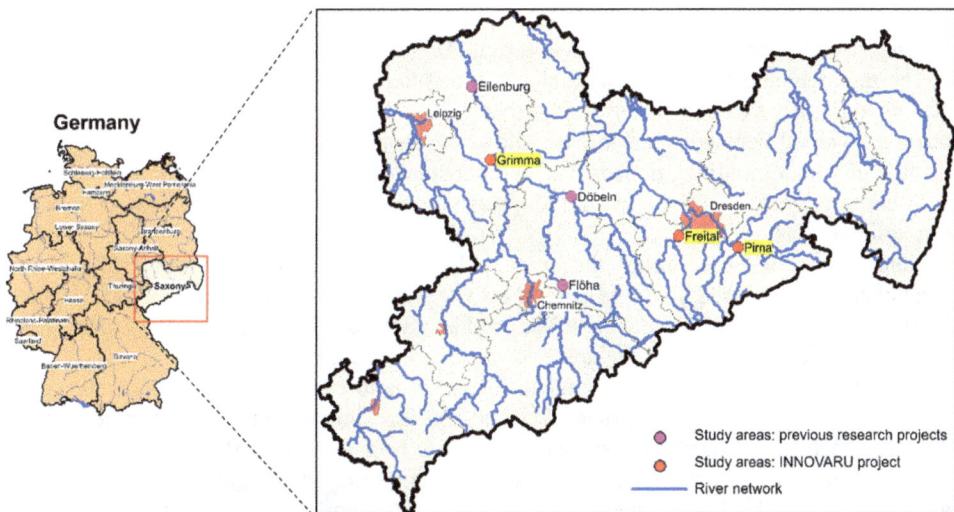

Figure 1: Study areas in the Free State of Saxony.

2.3 Flood scenarios

The improved methods in the paper will be validated against the damages of the 2002 flood in the study areas. The required inundation levels and flow velocities at the damaged buildings can be assigned (although only approximately) through a subsequent hydraulic simulation of the flood event. The influence of flow velocity on damage lead to the conclusion that very detailed 2D hydraulic models are necessary.

Such accurate models are available for the INNOVARU study areas of Freital, Grimma, Pirna and also for the town of Döbeln. For Eilenburg and Flöha, the flow around building stock is simplified considered over a mean roughness. The hydraulic calculations for the study areas of Grimma and Flöha are available from the Risk Management of Extreme Flood Events (RIMAX) project Methods for the Evaluation of Direct and Indirect Flood Losses (MEDIS) in which also the flow velocities for Eilenburg were derived [21]. The INNOVARU project comprises the re-interpretation of the 2002 flood event in the study area of Freital. 2D unsteady flow modelling was performed. For the 2002 flood in Pirna, a hydraulic calculation for the Elbe flood is available (Dam Authority Saxony).

In order to consider the upstream flooding from the tributaries Gottleuba and Seidewitz for the re-interpretation of the damage in the study area, the scenario HQ 100 (which shows a good match with the official inundation areas due to the 2002 flood) was assumed. Further information on the hydraulic models used within the INNOVARU project can be taken from Maiwald et al. [10]. In the course of further investigations, uncertainties considerations in the prognosis of the hydraulic action parameters would also have to be considered (cf. [22]).

3 VULNERABILITY OF BUILDINGS AND FLOOD DAMAGE

3.1 Damage scale for flooding

Based on the building stock and documented damage to buildings in August 2002, an initial five-stage differentiation of damage grades was implemented in Schwarz and Maiwald [23]; it represents one of the basic elements in the EDAC flood damage model [5].

The research on the vulnerability and risk assessment of buildings under extreme natural hazards in the sense of a multi-hazard approach [19], [20], also led to a further development into a six-graded damage scale in Maiwald and Schwarz [18] to include cases of severe damage in which buildings tipped over, were displaced from their foundations or washed away completely (cf. Table 1). In addition, washing out effects have to be considered in the description of damage grades.

Considering the observations during the March 2011 tsunami in Japan [15] and the tsunami damage analysed by EDAC after the 2010 Maule earthquake in Chile [24], a damage grade D6 (cf. Table 1) is introduced in Maiwald and Schwarz [18] in order to delineate these extreme cases of damage from the grade for common collapse (D5).

The introduction of D6 makes the identification of failure mechanisms comprehensive. In financial terms, both damage grades D5 and D6 are similar and represent a total loss. However, there are also differences that still need to be analysed. At the current state of the investigations, it can be assumed that additional demolition and disposal costs have to be taken into account for D5, but they do not apply for D6 if buildings are washed away.

3.2 Flood vulnerability classes

The concept of vulnerability classes was developed for determining earthquake intensity and observed shaking effects in the EMS-98 [17]. This concept was successfully transferred to flood hazard and damage interpretation by Maiwald and Schwarz [23] and further developed in Schwarz et al. [25].

Vulnerability classes subsume the event-specific vulnerability (or resistance) in the form of occurrence of similar damage grades (in quantity and quality) under the same impact intensities; they combine structures of comparable vulnerability in classes. The presented

Table 1: Enhanced flood damage scale (according to Maiwald and Schwarz [18]; an extension of Schwarz and Maiwald [23]).

Damage grade	Damage		Description
	Structural	Non-structural	
D1	None	Light	Only wetting through, dirt
D2	Light	Moderate	Slight cracking to loadbearing walls Doors/windows pushed in Washing out of foundations *Replacement of finishings necessary* *Contamination*
D3	Moderate	Heavy	Larger cracking in loadbearing walls and slabs Settlement Collapse ore of non-loadbearing walls *Replacement of non-loadbearing building elements necessary*
D4	Heavy	Very heavy	Collapse of loadbearing walls, slabs *Replacement of loadbearing walls, slabs*
D5	Very heavy	Very heavy	Collapse of larger parts of building
D6	Complete	Complete	Dislocation: building completely washed away, toppled or displaced from foundation

vulnerability tables for earthquakes [17], floods and wind [25] contain four vulnerability classes (A to D) for the typical building stock and two classes (E and F) for buildings specially designed to withstand the corresponding natural hazard. Flood vulnerability class HW-A (typical for clay buildings, cf. Table 2) represents the most vulnerable buildings with the largest expected flood impact damage; the buildings of vulnerability class HW-F (introduced in Schwarz et al. [25]) should survive a strong impact level without significant damage [10].

Table 2: Classification of building types into vulnerability classes [25].

Building type	Flood vulnerability class HW-					
	A	B	C	D	E	F
Clay	O					
Prefabricated timber frame	⊢O⊣	O	⊣			
Timber frame with masonry or clay infills	⊢O⊣	O	···⊣			
Masonry	⊢···	···O	···⊣			
Reinforced concrete			⊢···O			
Flood resistant design				⊢O		
Flood evasive design						O

O Most likely vulnerability class;
— Probable range;
·· Range of less probable, exceptional cases.

HW-E buildings usually consist of reinforced concrete or masonry structures and follow a flood resistant design; they are characterized by the separation of the main building from the flood impact (for instance, by raising the ground floor onto storey-high columns). Vulnerability class HW-F is assigned to flood evasive designed constructions like floating homes (cf. [26]), and structural solutions especially adapted to floods. These buildings, erected on pontoons, float when the water level rises and thus avoid flooding. Steel or concrete is mostly used for the pontoons. The construction of the actual building is of minor importance since contact with the flood is limited to the pontoon. Due to the lack of corresponding damage data, these constructions are not considered in the investigations.

The scatter and uncertainty of the appropriate vulnerability class(es) for each building type are described as follows: most likely vulnerability class, probable range and range of less probable, exceptional cases. The most likely class and the range of scatter result from a variety of damage assessments (cf. Table 2). The specific classification within the range of scatter depends on the condition and the structural design of the building.

3.3 Loss prediction

In contrast to conventional damage models (see, e.g., overview in Jongman et al. [1]), the vulnerability-relevant building parameters are also considered in the loss prognosis within the existing EDAC flood damage model. Following the proposed methodology, a set of specific damage functions (SDFs) for loss predictions were presented in Maiwald and Schwarz [5], [6], [13], [27]. Functions refer to the building type (SDF Type 1a) or flood vulnerability class (SDF Type 1b). A second type of functions (SDF Type 2) transfers the calculated damage grades D_i into losses.

The SDFs of the EDAC flood damage model could be successful validated based on the actual reported losses of the 2002 flood on residential buildings [12] in the study areas of Döbeln, Grimma and Flöha in Saxony [6], [27]. Additionally, also the results for the losses based on improved methods for the prognosis of the structural damage presented in Maiwald et al. [10] show a good agreement with the observed losses of the 2002 flood in all of the considered study areas [9]. In the current study, the functions of SDF Type 2 [27] are applied.

4 SIMULATIVE FLOOD DAMAGE MODELING AND VALIDATION

4.1 Elements of the uncertainty chain

Various elements of the uncertainty chain can be taken into account in a simulative flood damage modelling: uncertainties on the impact side (water level, flow velocity), uncertainties in the description of the existing structure (location, vulnerability class and replacement values), uncertainties in the development of structural damage (fragility functions) and the scatter in the losses. Based on the water level as impact parameter in Schwarz et al. [28] the damage modelling considers the simulation of the structural damage and a simulative variation of the location of the building. The present study concentrates on the uncertainties in the prognosis of structural damage and the resulting losses depending on water level and flow velocity.

4.2 Fragility functions for floods considering water level and flow velocity

Fragility functions depending on the water level over the ground floor level were presented in Schwarz et al. [28]. In the INNOVARU project, two improved types of fragility functions were derived depending on the water level over ground level h_{gl} and flow velocity v_{fl}. Flow velocities from the hydraulic calculations were assigned to the flood damage data.

The flow velocities of the tsunami at selected points were estimated in Suppasri et al. [29] and Foytong et al. [30] using video recordings. In Suppasri et al. [29] the differences between Plain coast and Ria coast were also highlighted. The estimated flow velocities in Suppasri et al. [29] and Foytong et al. [30] enable the derivation of average values of the Froude number F_r according to the coast classification in order to approximate flow velocities for the damage data (cf. eqn (1) and [19]).

$$v_{fl} = F_r \cdot \sqrt{g \cdot h_{gl}} . \qquad (1)$$

The term fragility functions (originating in earthquake engineering) means functions with which the probability of exceedance a certain damage grade can be determined depending on the magnitude of the impact [31]. In general, the cumulative logarithmic normal distribution (eqn (2)) is used to describe the functions mathematically.

$$F_{D_i}(x) = \Phi\left(\frac{\ln(x) - \mu}{\sigma}\right), \qquad (2)$$

where x = impact parameter ($x = h_{gl} \cdot v_{fl}^2$ or $x = h_{gl} + h_{gl} \cdot v_{fl}^2$); Φ = standard normal distribution; μ = logarithmic mean; σ = logarithmic standard deviation; $F_{Di}(x)$ is the conditional probability that the structure will reach or exceed the damage grade D_i, depending on the action parameter x. The parameters μ and σ are to be derived for each building type/ vulnerability class and damage grade. From eqn (3) the probability that a building will be damaged up to the damage grade D_i is calculated:

$$P[D_i \mid x] = F_{D_i}(x) - F_{D_{i+1}}(x). \qquad (3)$$

It should be noted that the scatter of damage results from the probabilities of occurrence of the individual damage grades according to eqn (3) and not by the distribution function according to eqn (2). The approach known from earthquake engineering has been applied to the natural hazard tsunami in Suppasri et al. [16], Maiwald and Schwarz [19], and Suppasri et al. [29], and also seems to be expedient for the flood effects considered here. In general, the particularities of flood impacts compared to tsunami impacts must be taken into account. In case of a tsunami, the flow velocity according to eqn (1) is linked to the water level h_{gl}. This is much more complex in the event of a flood. Floods can only show a slight to standing water movement, even with high water levels.

In **approach 1**, the momentum flux $x = h_{gl} \cdot v_{fl}^2$ is examined as an impact parameter. Here, for very low or non-existent flow velocities ($v_{fl} \approx 0$), despite possible larger water level h_{gl}, the action parameter x is also 0. In order to compensate for this, in **approach 2** a not true to the unit combination of the two action variables $x = h_{gl} + h_{gl} \cdot v_{fl}^2$ is chosen. This has the advantage that at low flow velocities v_{fl} the impact parameter x is determined by the water level h_{gl}, while the flow velocity becomes dominant in the case of a highly dynamic flood.

Fig. 2 shows the probabilities of exceedance $P(D > D_i)$ of the combined damage data set for the vulnerability class HW-C (typical for masonry buildings). The plausible transition between the flood and tsunami damage data is clearly visible. The fragility functions can be derived from the combined data set using a non-linear regression. Fig. 3 displays the fragility functions of approach 2. For vulnerability class HW-A, engineering assumptions were made to estimate the control parameters, since the data coverage is currently still too low.

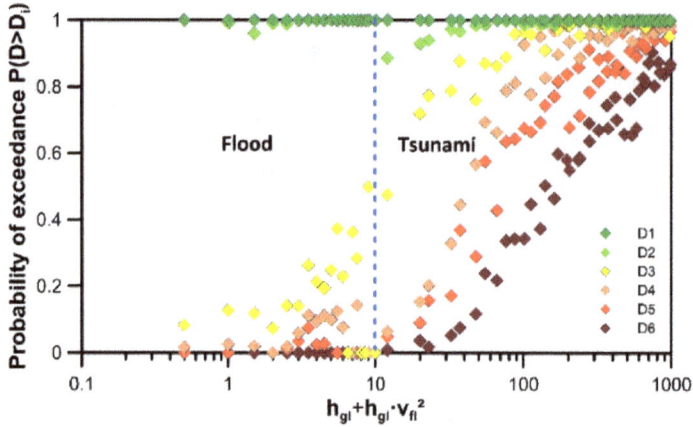

Figure 2: Probabilities of exceedance of the damage grades D_i for the vulnerability class HW-C (combined data set).

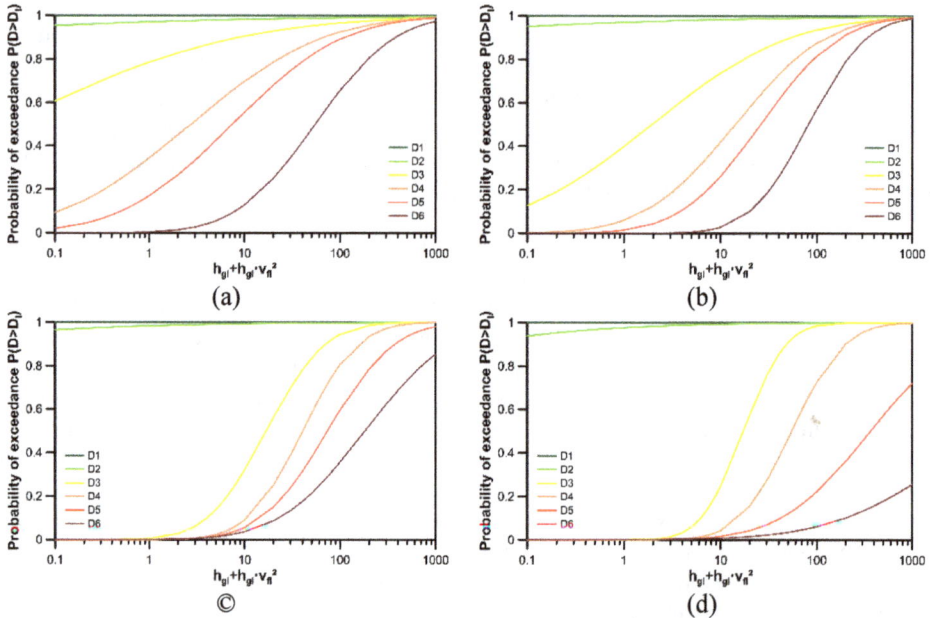

Figure 3: Fragility function of approach 2. (a) HW-A; (b) HW-B; (c) HW-C; and (d) HW-D.

4.3 Validation

The innovative options and advantages of the whole procedure are demonstrated by the case studies from the 2002 flood in Saxony. With the developed fragility functions, the probabilities of exceedance for the individual damage grades can be calculated for each individual building (microscale level) in the six study areas and towns of Saxony (Fig. 1).

The damage prognosis with the fragility functions also allows the transfer to financial loss indicators. A procedure for a Monte Carlo simulation was developed, which realises "n times" the damage grades according to the probability of exceedance according to the corresponding fragility functions. The losses are calculated for each of these damage grade realisations in the study areas, so that at the end n damage grade and loss scenarios are available.

Fig. 4 shows the comparison of the mean values of the losses for n = 1,000 simulations in comparison to the reported losses for the scenarios of the flood 2002. Both approaches show a good agreement with the actually reported losses for residential building stock (Fig. 4(a)), whereby the approach 2 results in a slightly better approximation. Underestimates (Eilenburg, Flöha, Freital, Grimma) but also overestimates (Döbeln, Pirna) of the real observed losses are possible here, so that there is no unique trend for the deviations.

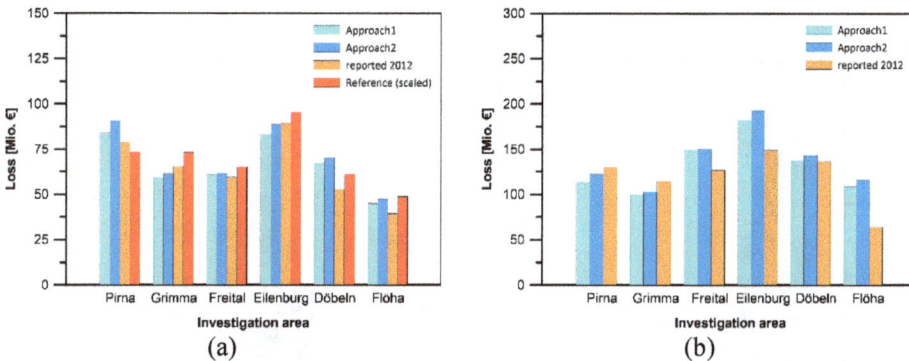

Figure 4:   Comparison of the mean values of the losses (n = 1000 simulations) and the reported losses HW 2002 [12]. (a) Residential buildings; and (b) Entire building stock. Reported 2012: Final loss statement from the SAB [12]. Reference (scaled): Loss scaled to the number of residential buildings actually considered.

In areas with large industrial complexes (Eilenburg, Flöha), both approaches lead to a significant overestimation of the losses for the entire building stock (Fig. 4(b)). Individual case analyses would be necessary for such industrial complexes. But overall, the calculations can be approximated well to the losses that actually occurred. It can be stated that the newly developed method also delivers reliable results with regard to the prognosis of the expected values of the losses. A further improvement in the quality of the loss prognosis and a refinement in the consideration of the building parameters can be expected in future by application of the new developed building typological synthetic damage functions in Golz et al. [11].

The inconsistent trend in the deviations indicates the importance and the need to consider the uncertainties in the flood damage prognosis, in more detail. Fig. 5 shows the scatter of the losses for the residential building stock, which follows from n = 1000 Monte Carlo simulations according to approach 2. The mean value of the losses (cf. Fig. 4(a)), the ± 1σ standard deviation and also the reported losses of the SAB [12] are marked. (Note: For Grimma, actual losses are slightly outside the range shown.) For the residential building stock, the calculated standard deviations can be assessed as relatively low in the range of 1.0%–2.1%, while the overall range of scatter in the losses is between 8.2% and 11.6%.

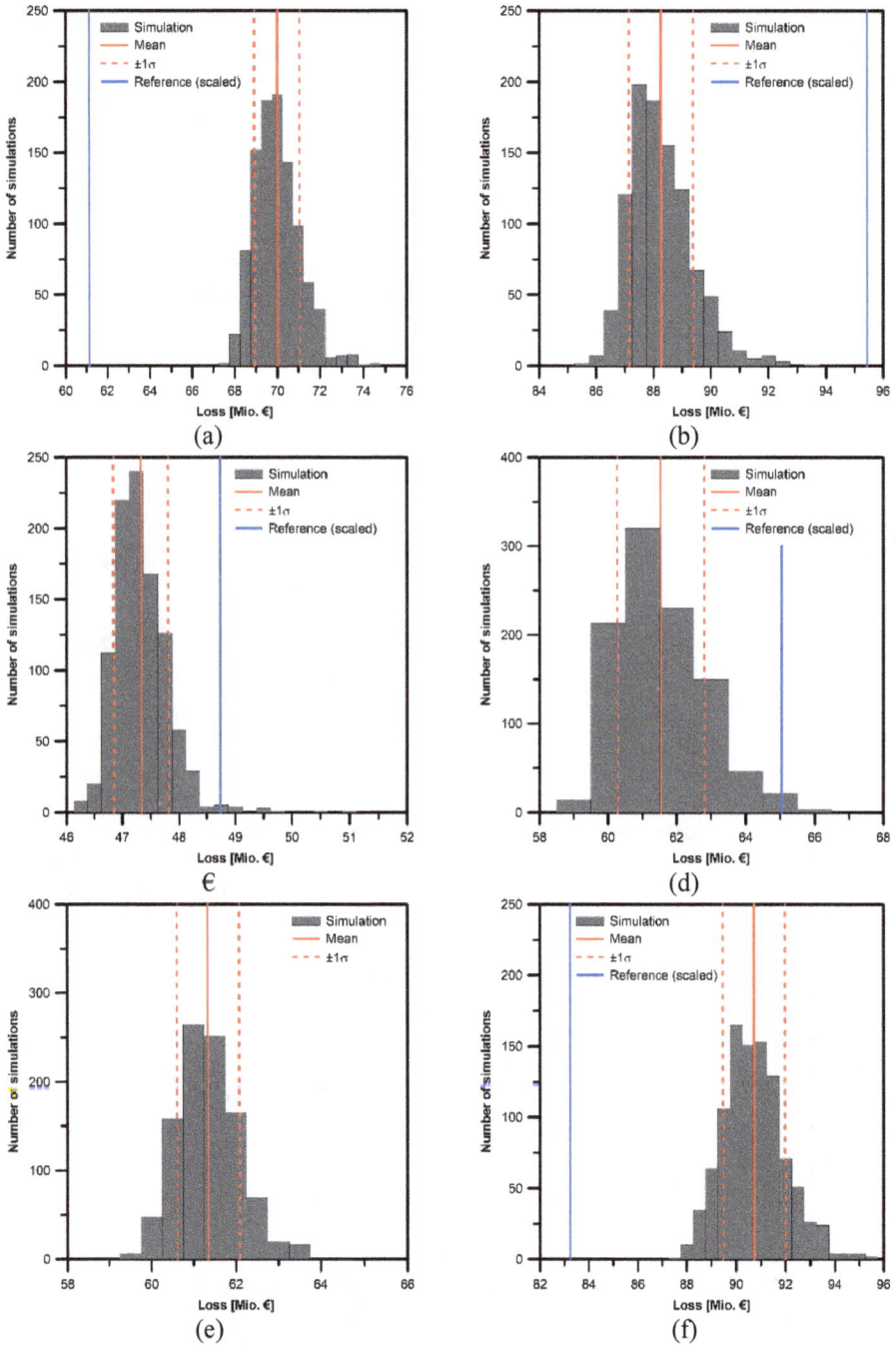

Figure 5: Calculated scatter of losses for residential building stock in the study areas (2002 flood) according to approach 2. (a) Döbeln; (b) Eilenburg; (c) Flöha); (d) Freital; (e) Grimma; and (f) Pirna.

Although the actually observed losses are mostly slightly outside of the calculated scatter range, it should be noted that the absolute deviation of the mean values is only approximately 10% on average, which is similar to the overall range of scatter in the losses. It has to be emphasized that these studies relate to one element of the uncertainty chain, only.

For this reason, other factors must also be taken into account in the future when describing the wide scatter of flood damage. The uncertainties or scatter that are "lost" when summarizing the range of variation which are related to the concrete construction/geometry and replacement value of the buildings and also the uncertainties which are related to the hydraulic scenarios [22] have not yet been considered. Other influencing factors have a random influence on the development of the damage (aleatoric uncertainties), but for technical and practical reasons they cannot be represented exactly in the prognosis model at present. These include, for example, the impact of debris, scouring and foundation erosion as well as unspecified contamination or pollution. The authors are convinced that the derivation and determination of certain probabilities of occurrence in connection with a Monte Carlo simulation offers possibilities for future consideration of these type of uncertainties.

5 CONCLUSIONS AND OUTLOOK

Considering the elements of the existing EDAC flood damage model, new approaches for the prognosis of the structural damages due to flood impact are developed in the paper. The presented model provides basic tools for the prediction of structural damage of the building including the scatter taking into account the specific building vulnerability, the inundation level and the flow velocity. The developed fragility functions represent an element for the consideration and identification of uncertainties in the flood damage and loss prognosis.

In this paper, a realistic re-interpretation of the real observed damages from the 2002 flood in Saxony is carried out. Results are presented for six different study areas with moderate flow velocities. Using the Monte Carlo method, the presented fragility functions enable a simulative prognosis of structural damage of the single buildings due to flood impact considering water level and flow velocity with two different approaches. The damage grades for n = 1,000 Monte Carlo simulations are calculated for each study area and converted into loss statements using the existing damage functions of the EDAC flood damage model. In addition to the expected values of the losses, the scatter range and uncertainties are also identified.

In all cases, a remarkably good agreement between the predicted and the reported losses can be stated for inventory of residential buildings, whereby approach 2 results in a slightly better approximation. A further improvement in the loss prognosis can be expected by the application of new synthetic damage functions that take into account the concept and engineered classification of the damage grades (as derived in the INNOVARU project [11]).

The derived new fragility functions enable in principle also damage prognosis for flash flood events with high flow velocities. Current investigations into the damage caused by the 2021 flood in the Ahr Valley in western Germany [32] are preparing the validation of the new approaches to such flood events.

With the fragility functions, a first element for considering the uncertainty chain is available, which initially lead to a relatively small scatter range in the loss prognosis. As target for the next step, the uncertainties of the building parameters, of the losses by the same damage grade, in the description of the impact parameter (water level, flow velocities, duration) and also aleatoric uncertainties would have to be considered. Taking into account the entire chain of uncertainty would offer new application possibilities in the context of cost–benefit studies. The previous approach in Germany, that cost/benefit ratios > 1 leads to the rejection of the measure, could be interpreted more generously in the future, for example if the costs of the measure lie within a scatter range of the damage prognosis that has to be defined.

ACKNOWLEDGEMENT

Significant parts of the new methodological approaches presented in this paper were elaborated as part of the joint research project funded by the German Federal Ministry of Education and Research (BMBF): Innovative Vulnerabilitäts- und Risikobewertung urbaner Räume gegenüber Überflutungsereignissen (INNOVARU; English translation "Innovative Vulnerability and Risk Assessment of Urban Areas Against Flood Events") under grant agreement FKZ:13N14931.

REFERENCES

[1] Jongman, B., Kreibich, H., Apel, H., Barredo, J.I., Bates, P.D., Feyen, L., Gericke, A. Neal, J., Aerts, J.C.J.H. & Ward P.J., Comparative flood damage model assessment: Towards a European approach. *Natural Hazards and Earth System Sciences*, **12**(12), pp. 3733–3752, 2012.
[2] Black, R.D., Floodproofing rural residences. Washington D.C. Report no EDA 77-088. US Department of Commerce, Economic Development Administration, 1975.
[3] Sangrey, D.A., Murphy, P.J. & Nieber, J.L., Evaluating the impact of structurally interrupted flood plain flows. Technical Report No. 98, Cornell University Water Resources and Marine Sciences Center, Ithaca, New York, USA, 1975.
[4] Clausen, L. & Clark, P.B., The development of criteria for predicting dambreak flood damages using modelling of historical dam failures. *International Conference on River Flood Hydraulics,* John Wiley & Sons Ltd, 1990.
[5] Maiwald, H. & Schwarz, J., Ermittlung von Hochwasserschäden unter Berücksichtitung der Bauwerksverletzbarkeit, EDAC-Hochwasserschadensmodell, Scientific Technical Reports 01-11, Universitätsverlag, Bauhaus-Universität Weimar, 2011.
[6] Maiwald, H. & Schwarz, J., Damage and loss prognosis tools correlating flood action and building's resistance-type parameters. *International Journal of Safety and Security Engineering*, **5**(3), pp. 222–250, 2015.
[7] Naumann, T. & Rubin, C., Ermittlung potenzieller Hochwasserschäden in Pirna nach dem gebäudetypologischen VERIS Elbe-Ansatz. *Tagungsband zum DWA-Seminar Hochwasserschadensinformationen: Neues und Bewährtes*, Hennef, pp. 86–101, 2008.
[8] Naumann, T., Golz, S. & Nikolowski, J., Synthetic depth-damage functions: A detailed tool for analyzing flood resilience of building types. *Final Conference of the COST action C22 Urban Flood Management in cooperation with UNESCO-IHP*, Paris, 2009.
[9] Maiwald, H., Kaufmann, C., Langhammer, T. & Schwarz, J., A new model for consideration of flow velocity in flood damage and loss prognosis. *FLOODrisk 2020 – 4th European Conference on Flood Risk Management*, Paper 11_9, 2021.
[10] Maiwald, H., Schwarz, J., Kaufmann, C., Langhammer, T., Golz, S. & Wehner, T., Innovative vulnerability and risk assessment of urban areas against flood events: Part 1 – Prognosis of structural damage with a new approach considering the flow velocity. *Natural Hazards* (accepted for publication), 2022.
[11] Golz, S., Maiwald, H., Naumann, T., Schwarz, J., Innovative vulnerability and risk assessment of urban areas against flood events: Part 2 – Prognosis of losses with a new approach for synthetic damage functions. *Natural Hazards* (forthcoming), 2022.

[12] Sächsische Aufbaubank/Saxony Relief Bank (SAB), Information to the losses on the residential and commercial building sector due to 2002 flood, submitted in December 2012.

[13] Maiwald, H. & Schwarz, J., Berücksichtigung der Fließgeschwindigkeit bei Hochwasser-Schadensmodellen. *Bautechnik*, **86**(9), pp. 550–565, 2009.

[14] Ministry of Land, Infrastructure and Transportation, Survey of tsunami damage condition. http://www.mlit.go.jp/toshi/toshi-hukkou-arkaibu.html. Accessed on: 2 Feb. 2022.

[15] Suppasri, A., Mas, E., Charvet, I., Gunasekera, R., Imai, K., Fukutani, Y., Abe, Y. & Imamura, F., Building damage characteristics based on surveyed data and fragility curves of the 2011 Great East Japan Tsunami. *Natural Hazards*, **66**(2), pp. 319–341, 2013.

[16] Suppasri, A., Charvet, I., Imai, K. & Imamura, F., Fragility curves based on data from the 2011 Tohoku-Oki Tsunami in Ishinomaki City, with discussion of parameters influencing building damage. *Earthquake Spectra*, **31**(2), pp. 841–868, 2015.

[17] Grünthal, G., Musson, R., Schwarz, J. & Stucchi, M., European macroseismic scale 1998. *Cahiers du Centre Européen de Geodynamique et de Seismologie*, ed. G. Grünthal, **15**, Luxembourg, 1998.

[18] Maiwald, H. & Schwarz, J., Unified damage description and risk assessment of buildings under extreme natural hazards. *European Journal of Masonry*, **23**(2), pp. 95–111, 2019.

[19] Maiwald, H. & Schwarz, J., Vulnerability assessment of multi hazard exposed building types: Development of an EMS-98 based empirical-statistical methodology. *16th World Conference on Earthquake Engineering,* Santiago, Chile, Paper No. 2134, 2017.

[20] Schwarz, J., Maiwald, H., Kaufmann, Ch., Langhammer, T. & Beinersdorf, S., Conceptual basics and tools to assess the multi hazard vulnerability of existing buildings. *European Journal of Masonry*, **23**(4), pp. 246–264, 2019.

[21] Kreibich, H., Piroth, K., Seifert, I., Maiwald, H., Kunert, U., Schwarz, J., Merz, B. & Thieken, A., Is flow velocity a significant parameter in flood damage modelling? *Natural Hazards and Earth System Sciences*, **9**(5), pp. 1679–1692, 2009.

[22] Bhola, P.K., Leandro, J., Konnerth, I., Amin, K. & Disse, M., Dynamic flood inundation forecast for the city of Kulmbach using offline two-dimensional hydrodynamic models. *13th International Conference on Hydroinformatic*, Palermo, Italy, 2018.

[23] Schwarz, J. & Maiwald, H., Berücksichtigung struktureller Schäden unter Hochwassereinwirkung. *Bautechnik*, **84**(7), pp. 450–464, 2007.

[24] Maiwald, H., Schwarz, J., Abrahamczyk, L. & Lobos, D., Das Magnitude 8.8 Maule (Chile)-Erdbeben vom 27. Februar 2010 – Ingenieuranalyse der Tsunamischäden. *Bautechnik*, **87**(10), pp. 614–622, 2010.

[25] Schwarz, J., Maiwald, H., Kaufmann, C. & Beinersdorf, S., Evaluation of the vulnerability of existing building stocks under single and multi-hazard impact. *16th European Conference on Earthquake Engineering (ECEE)*, Thessaloniki, Greece, 18–21 Jun., Paper 11641, 2018.

[26] Strangefield, P. & Stopp, H., Floating houses: An adaptation strategy for flood preparedness in times of global change. *Flood Recovery, Innovation and Response IV*, 2014.

[27] Maiwald, H. & Schwarz, J., Schadensmodelle für extreme Hochwasser – Teil 1: Modell-bildung und Validierung am Hochwasser 2002. *Bautechnik,* **91**(3), pp. 200–210, 2014.

[28] Schwarz, J., Maiwald, H. & Kaufmann, C., Unsicherheiten bei der Quantifizierung von Hochwasser-Schadenspotenzialen. *Bautechnik*, **93**(4), pp. 214–229, 2016.

[29] Suppasri, A., Imai, K., Imamura, F. & Koshimura, S., Comparison of casualty and building damage between Sanriku Ria Coast and Sendai Plain Coast based on the 2011 Great East Japan Tsunami. *International Sessions in Conference of Coastal Engineering, JSCE*, **3**, pp. 76–80, 2012.

[30] Foytong, P., Ruangrassamee, A., Shoji, G., Hiraki, Y. & Ezura, Y., Analysis of tsunami flow velocities during the March 2011 Tohoku, Japan, Tsunami. *Earthquake Spectra*, **29**(1), pp. 161–181, 2013.

[31] Hazus®-MH MR5, Earthquake loss estimation methodology, Advanced Engineering Building Module (AEBM), Technical and User's Manual, Federal Emergency Management Agency. https://www.hsdl.org/?view&did=12760. Accessed on: 5 Mar. 2022.

[32] Hochwasser in Rheinland-Pfalz und Nordrheinwestfalen 2021, Earthquake Damage Analysis Center, Weimar. https://edac.biz/forschung/hochwasser/feldeinsaetze/hochwasser-rheinland-pfalz-und-nordrheinwestfalen-2021. Accessed on: 21 Mar. 2022.

SECTION 2
FLOOD MITIGATION

QUARENTA STREAM RENATURALIZATION: HARMONY BETWEEN DEVELOPMENT AND HISTORICAL PRESERVATION IN BRAZIL

ADELINA CRISTINA AUGUSTO CHAVES[1], BRENDO BENEDITO DE SOUZA[1]
& JUSSARA SOCORRO CURY MACIEL[1,2]
[1]Federal Center for Technological Education of Amazonas, Brazil
[2]Geological Survey of Brazil, Brazil

ABSTRACT

Climatic issues have drawn the attention of researchers to the Amazon, referring above all to environmental sustainability and the preservation of the largest tropical forest in the world. However, urban environmental degradation is related to the stagnation of watersheds by canalization; rectification of watercourses and urban spaces occupation disorderly, impacted by social inequalities, compromise socio-environmental and cultural sustainability. Avoiding or mitigating such degradation is generally attributed to the sphere of public policies, whose insufficiency and/or absence aggravate urban problems, whether water and sewage, atmospheric pollution, solid and/or industrial waste. Manaus, the largest financial, corporate, and commercial point in the north of Brazil is embedded in the heart of the Amazon Forest, endowed with a set of hydrographic basins, whose numerous streams permeate all areas of the city, coexists each year with the recurring Negro River floods. In 2021, the Rio Negro flood reached 30.02 m, a historic record in 119 years. The official data indicate that in that year 15 districts suffered flooding, impacting the lives of approximately 24,000 people. The Educandos district, one of the oldest and with the greatest urban concentration, whose history is intertwined with the history of Manaus, located in the middle-south of the city, was very affected by the flood. The mainsprings of the watershed that permeates Educandos are the Educandos, Mestre Chico, and Quarenta streams, which flow into the Negro River. This work, based on the successful experience of a Mindu section basin renaturalization and the creation of the Mindu National Park, studies the important and urgent need for the Quarenta stream renaturalization, to prevent and minimize floods in Manaus, while proposing the creation of a historical site of stilts and the floating city, preserving the culture and respect for the native people.
Keywords: urban floods, renaturalization, Quarenta stream.

1 INTRODUCTION

Climate issues have drawn the attention of researchers to the Amazon, referring above all to environmental sustainability and the preservation of the world's largest tropical forest, where record droughts and floods are an unmistakable indication of ongoing climate change. The Amazon River, located in South America, is the largest river in terms of water volume in the world and the second largest in terms of land area [1]. With about 6,992 km, it travels the north of South America, the Amazon Forest, and flows into the Atlantic Ocean [2]. Among its more than a thousand tributaries are the rivers: Madeira, Negro, and Japurá. Which are also among the ten largest rivers on the planet [3]. The Amazon, with the largest hydrographic basin in the world, with more than 7,000,000 km², is responsible for about one-fifth of the total river flow in the world [4], with the water that flows through Amazon rivers equivalent to 20% of the water liquid sweet from Earth [3].

The Amazon River has several names, since its origin at the source of the Apurímac River (upper western part of the Andes Mountain range), in southern Peru, enters Brazilian territory under the name of the Solimões River and finally, in Manaus, after Negro River junction, it receives the name of Amazonas and as such follows until its mouth in the Atlantic Ocean. Since 1902, researchers have been monitoring the water level of the Rio Negro, measured as

WIT Transactions on The Built Environment, Vol 208, © 2022 WIT Press
www.witpress.com, ISSN 1743-3509 (on-line)
doi:10.2495/FRIAR220041

it passes through the port of Manaus. The study results, published in Science Advances, show that extreme events are increasingly frequent. Droughts and floods, until the middle of the last century, occurred almost in parallel and had a frequency of twenty years. Now, floods occur every four years.

The three largest floods recorded at the Port of Manaus: 30.02 m; 29.97 m and 29.77 m, occurred in the last 12 years (2021, 2012, and 2009, respectively) [5]. In Manaus, the announced tragedy causes flooding, where houses on the banks of the streams are invaded by water, causing loss of goods, a proliferation of diseases caused by contaminated water, interruption of essential services, and widespread unhealthiness for the affected population. The contaminated water stench is involved with garbage, human waste, and rats that float around the houses. In addition to snakes, mosquitoes, and other poisonous and/or disease-transmitting animals, they leave the local population in a situation of extreme vulnerability.

In Manaus, public policies have focused on emergency actions, seeking to help the population through the construction of temporary bridges, made of wooden slats to give mobility to the affected population, obstruction of manholes, removal of garbage from streams, distribution of baskets basic and rent aid. However, bold public infrastructure policies that make it possible for the population to coexist with the hydrological cycle of the Rio Negro are still far from what is necessary. Despite the successful Social and Environmental Program of Igarapés de Manaus (PROSAMIM) and the successful restoration action of the Mindu basin piece that culminated in the creation of the Mindu Park.

In 2021, the Negro River flood reached 30.02m, a historic record in 119 years of a historical series. Manaus is a city embedded by the Rio Negro, with around 150 polluted streams and families living on its banks. Official data indicate that, in that year, 15 districts suffered flooding, impacting the lives of approximately 24,000 people. Educandos, located in the south center of the Manaus, was one of the hardest hit by the floods.

In this work, based on the successful experience of restoration in the Mindu basin, we propose the renaturalization of the Quarenta stream, at its mouth in the Educandos district, restoring banks and integrating elements of historical and cultural preservation.

2 RESTORATION OF WATERCOURSES IN URBANIZED AREAS

Restoration is the process of a place intentional alteration to its natural form through processes and processes that lead to the reestablishment of the sustainability relationship and health between the natural and the cultural, as defined by the Ecological Restoration Society [5], [6]. The goal, according to scholars, is to simulate the structure, function, diversity, and dynamics of a specific ecosystem, according to its historical characteristics [6]–[9]. However, given the difficulty or even impossibility of returning to the original ecosystem, the authors have considered expanding this concept, incorporating other dimensions such as landscape, ecological, and water quality within the perspective of river restoration [7]. Research in Saxony-Anhalt (Germany) highlight the special importance of integrated ecological methods in measuring the success of streams and rivers renaturation [10].

Scientists recognize this research agenda is diverse and ambitious. But recommend we should also not ignore the existing knowledge that can already be lever aged, building on the practice changes that have been achieved through efforts to implement water-sensitive cities [11]. However, the solutions implemented are complex, including technological and social innovations, urban reorganization, public policies, cooperation between different sectors of society, socio-environmental education, behavior change.

In Brazil, restoration/renaturalization terms are used, without distinction [12], [13]. Thus, in this work, they mean human interventions in all or parts of watercourses, with a view to the environmental recovery of the function, balance, and dynamic but the sustainable ecosystem behavior. This also includes removing any disturbance so that the natural process can recover. This is not necessarily about recovering a pre-existing system, but about restoring landscapes that are stable from a hydrological and geomorphological point of view, capable of supporting healthy and biodiverse ecosystems, while preventing urban floods.

3 THE SUCCESSFUL RESTORATION EXPERIENCE OF MINDU, MANAUS, AM

The RIMA Environmental Impact Report [14] prepared to support the Mindu restoration project, considers the Igarapé do Mindu to be the most expressive watercourse in the urban area of Manaus, capital of the State of Amazonas, whose basin According to the aforementioned report, it occupies approximately 1/4 of the urban territory and concentrates approximately 30% of the population. It also reports that the disorderly occupation of part of the basin over the decades resulted in a series of environmental and social problems, enabling unhealthy conditions for its residents. It points out among the main consequences, polluted streams and watercourses, silted up and obstructed both by the clearing of vegetation and by the presence of large amounts of garbage, resulting from the construction of houses on its banks, and in some cases, in the stream bed itself.

Therefore, Mindu restoration required a set of socio-environmental and engineering interventions, such as sanitation works and services; urbanization; Housing; improvement of the road system, and environmental protection in the area of influence of the streams located in the Mindu Hydrographic Basin. The project is part of a Federal Government Program entitled "Sanitation for All", it has as a proponent the Municipality of Manaus in partnership with Federal Bank (Caixa Econômica), within the Stormwater Management modality, under contract No. 216,881, with a budgeted amount in 120 million of Reais, of which 108 million correspond to financing and 12 million as counterpart. The Resettlement Plan, as a result of the implementation of engineering works expropriations, it will contemplate the acquisition of properties for 2,285 families, residing in Permanent Preservation Areas (APP) of Igarapé do Mindu, distributed in the communities that make up the district. Each family consists of an average of 5 people, making a total of approximately 11,425 people, to be benefited. The Mindu creek (Fig. 1), the scope of the Manaus City Hall revitalization project, is located in the large hydrographic basin of the São Raimundo creek. At its mouth is the Franceses stream, forming the Cachoeira Grande stream, which flows into the São Raimundo watercourse (Fig. 2).

The hydrographic basin, the physical base of hydrological studies, is an area of natural precipitation capture that makes the flow converge to a single exit point, the outlet. The water displacement on the basin surface, in rivers, canals and reservoirs are one of the most important parts of the hydrological cycle. In the aforementioned report, the hydrographic basin of the Mindu creek has the Cachoeira Grande creek as its outlet. The set of interventions was varied, according to the section covered: the Ecological Corridor – did not suffer major interventions, as it is a protected area (green area) of the city.

The Mindu river has 21 km in its entirety, where the corridor corresponds to 7.0 linear km, therefore the intervention area covered an area of 14 km (before and after the corridor), this section starts at the source of the same. igarapé, in the Cidade de Deus district, close to the Adolpho Ducke Forest Reserve, following the Colina do Aleixo bridge. At this point, the Ecological Corridor begins (an area that has not suffered significant interference), ending at Av. Humberto Calderaro Filho. Where the second stretch of interference began, ending at its mouth, located near the Ponte dos Bilhares, in the Nossa Senhora das Graças district.

Figure 1: Mindu stream, from its source to its mouth [14].

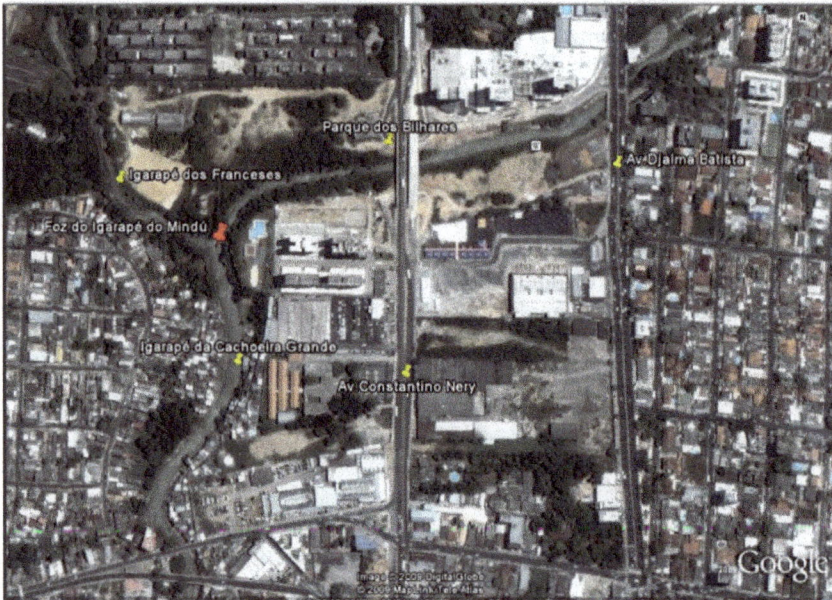

Figure 2:   The mouth of the Mindu stream, confluence with the French stream and the Cachoeira Grande stream [14].

Therefore, the intervention section of the Mindu stream revitalization project; starts at its source, in the Cidade de Deus district, close to the Adolpho Ducke Forest Reserve, it is interrupted by urban occupation, in the districts: São José; Armando Mendes; Mutirão and other districts in the east side of Manaus, returning close to the SESI Worker's Club, passing through private green areas (Private Natural Heritage Reserve – Honda) and green areas of housing developments (Villar Câmara, Tiradentes, Petros – on the left bank – Barra Bela,

Jardim Primavera, Imperial Park – on the right bank). The corridor brings together four Municipal Conservation Units: the Adolpho Ducke Botanical Garden, the Nascentes do Mindu Municipal Park, the Mindu Municipal Park, a Private Natural Heritage Reserve owned by Honda. The entire course of the Mindu stream was revitalized.

To support the recovery planning of the river banks, a situation and zoning survey of their occupation was carried out. This zoning defines different strategies for margin recovery, cf. the occupation zone: relocation of existing houses, in areas of high population density, implementation of Linear Parks in a marginal strip of the watercourse and a preservation strip revegetation, in addition to integration between vegetation and public spaces such as the implementation of lanes walk, bike path, gym equipment, and children's playgrounds, giving the population opportunity to enjoy these environments and supervise the conservation of watercourses. The areas that still have a riparian forest for a large part of their extension were defined as Urban Ecological Corridor in which the vegetation located in the Permanent Preservation Area allows connectivity between different already existing protected areas. In areas of relevant environmental or urban interest, conservation units or public parks were implemented.

Among the main works carried out for the renaturalization of Mindu stand out:

- Engineering and Architecture: Execution of macro and micro-drainage, with reservoirs implementation (or swimming pools) and restoration of natural gutters; Collection and Transport of Civil Construction Waste; The Road System Expansion; Urban Requalification and Landscaping, with the main results – the environmental recovery of its margins; containment implementation works and creation of natural water retention spaces. In addition, the reincorporation of the watercourse into the urban landscape with social and urban requalification along its path.
- Social: Relocation of floodplain residents; insertion of public leisure spaces such as small sports courts, children's playgrounds, and bicycle paths; The linear park was bordered by a right of way for infrastructure for basic services, with public roads to facilitate the movement of the local population.
- Sanitary: Concentration of actions such as garbage and sediments removal.
- Ecological: The creation of the Municipal Park of the source Mindu River, and Municipal Conservation Unit. Existing private spaces in the Park area were expropriated by the City Hall; implementation of the Mindu Stream Ecological Corridor, which allows, through the preserved Permanent Preservation Area, connectivity between the Mindu Municipal Park, the Honda Natural Heritage Private Reserve, and the various green areas of the surrounding housing complexes; the expansion of the Bilhares' Bridge Park (located along the Mindu) to the São Jorge bridge, after the relocation of the existing houses in the permanent preservation strip of the Cachoeira Grand stream.
- Technicians Studies: Preparation of Complementary Master Plans for Drainage, Sanitation, Protection of the Watercourse Margins, to provide basic information to municipal managers in the environmental, urban, and infrastructure areas of Manaus, on the best alternatives from the economic and socio-environmental point of view for the management of urban waters in the studied area.

The restoration of the Mindu resulted in recovered riverbanks and springs; reduced flooding effects, with macro-drainage and clearance of streams; urbanized margins with structures for recreation and leisure and a low level of waterproofing, due to the implantation of green areas; new roads implemented and interconnection of existing ones, as part of the proposed implementation of the new road system and facilitated circulation in the adjacent road network; the public reduction spending in the fight against endemic water transmission; the generation of work and income during construction; the optimization of the result of infrastructure maintenance actions, particularly macro-drainage; the improvement of urban mobility, with the creation of a cycle path for alternative transport linking the East Zone to the South Center Zone of the city, reducing the pressure on the collective public transport system and illegal means of transport, in addition to the elaboration of Complementary Plans to the Manaus Urban and Environmental Master Plan, specifically: Master Plan for Drainage, Sanitation, and Protection of the banks of the Watercourses.

4 RENATURALIZATION OF IGARAPÉ DO QUARENTA: HARMONY BETWEEN HISTORICAL PRESERVATION AND DEVELOPMENT

In 2021, the Rio Negro flood reached 30.02 m, a historic record in 119 years. Official data indicate that, in that year, 15 districts suffered flooding, impacting the lives of approximately 24,000 people. Educandos district, located in the south center of the city, was one of the oldest and with the greatest urban concentration, whose history is intertwined with the own history of Manaus, was one of the most affected districts.

The mainsprings of the watershed that permeates Educandos are the following streams: Educandos, Mestre Chico, and Quarenta, which flow into the Negro River. Educandos, in the South Center Zone of Manaus, is one of the oldest and with the greatest urban concentration, with commercial and industrial areas. Its perimeter starts on the Negro River between the Educandos stream and the 40 streams; goes to Leopoldo Peres Avenue and goes to Pres. Kennedy Avenue to the source of the Colonia Oliveira Machado stream; returning to the Negro River to the Educandos stream.

The history of the Quarenta/Educandos basin, represented in Fig. 3, is intertwined with the own history of Manaus. It goes back to the rubber boom, whose economic effervescence boosted trade, with an intense flow of migrants (1), which gave rise to the floating city (2).

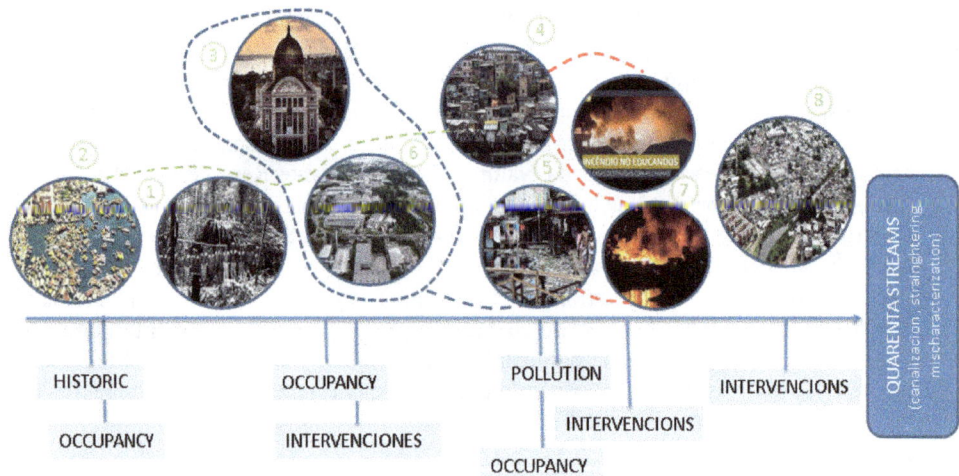

Figure 3:   History of the Educando district, where the exit of the Quarenta stream is located. *(Source: Adapted by the author. Images www.acritica.com.br.)*

The wealth generated by the rubber cycle brought to Manaus an urban model mirrored in the French model, known as the Belle Epoque (3), where cities were seen as chessboards with their aligned and parallel streets, which resulted in the removal of original housing and regulations that prevented certain types of construction in downtown. Thus, the original populations were removed (4). Another important consequence was the canalization of rivers and streams of the downtown, with the mere function of sewage collectors being delegated to the other remaining streams. The population that was being displaced, tried to remain as close as possible to their original habitat, causing the intense occupation of the remaining streams' banks (5). This occupation and the disorderly growth of Manaus, in more recent history, was also intensified with the creation of the Manaus Free Trade Zone (6). The absence of public policies and/or the State's indifference to those populations further aggravated the health, education, housing, and security issues that generated numerous problems, including the large fires (7), which marked the basin's history of the Quarenta/Educandos. Recently, an important intervention, the PROSAMIM (8) project, which, despite having brought great and significant improvements to the region and to a part of the population that managed to remain in place, adopted, centuries later, the same solution for watercourses, the channelling, and mischaracterization of the landscape.

Another emblematic issue in the struggle for survival at the mercy of the waters can be traced back to the end of the rubber monopoly; the twenties of the last century economic crisis and; the demographic growth due to the migratory flow of riverside dwellers and workers from other regions to Manaus, aggravated by the scarcity of resources, contributed to the crisis of homelessness in Manaus. In this adverse scenario, in 1920, the floating city of Manaus emerged, as a result of the public policies absence aimed at poor populations and investments destined for the construction of low-income housing.

Initially, the floating city in its primitive version was a conglomeration of wooden houses covered with Buçu straw, light, and resistant fiber. The entire structure of the house, comprising a roof, walls, floor, and balcony, was placed on wooden logs seated on Açacu "floater", which ensured the security, functionality, and floating condition of the house. The streets of the floating city were made with huge and thick kapok planks, the widths varied between about 50 cm [15] There was everything a city needs for its existence: pharmacy, nightclub, clothes, shoes, stores everything the river gave this. The city served as an entrepot for large companies that exported alligator hides, rubber, pirarucu, jaguar skins, snakes, gold, and contraband. The Floating City gained features in national magazines and international notoriety. However, the floating dwellings did not present the minimum conditions of comfort and hygiene for their users and constituted a serious social problem.

The floating city extended towards the waters of the Manaus, Quarenta, and Cachoeirinha streams, reaching gigantic proportions [13]. In 1960, mon. official census data, Manaus had a population of 175,343 inhabitants. That year the floating city had 2,145 houseboats. "The incredible territorial and demographic dimensions of the floating Manaus ships that transform them into an urban phenomenon and a social issue for the authorities, in which the title: 'floating city', built during this time, is the most emblematic historical index. Faced with this new reality, some official surveys began to be carried out on the 'floating city', to better understand the 'problem' and then give an account of its 'resolution' materialized in its complete destruction in 1967" [16].

In 1967, Governor Reis (1964–1967), under strong pressure from the Amazon Military Command and the Port Authority, backed by the military government, dismantled the floating city, transferring its residents to several districts, among which we highlight Alvorada, Coroado, Costa e Silva and Santo Antônio Complex.

The dramatic outcome, however, did not destroy the Amazonian way of life, where it is common to find, until today, part of the riverside populations throughout the state living on their floating ships. The government itself, when the floods of the Negro River invade Manaus, making traffic unfeasible, resort to "floating fairs" so those producers can sell their products. In other words, the force of the way of life remains active in the region, being one of the most significant cultural expressions of the Amazonian man, inhabiting the waters and seeking their livelihood and leisure in them.

In this way, "the floating culture is one of the most significant and impactful ways of claiming the city to Manaus residents" [16].

The Quarenta stream is widely studied for its length (38 km), average width (6 m), and average depth (50 cm), as it has several springs. But mainly due to its socio-environmental degradation, chemical contamination, and accumulation of solid waste. The disorderly occupation, in the Educandos district, proved to be a favorable scenario for the occurrence of large floods, with a history of repetitions, accumulating significant environmental and social damage: houses collapse; proliferation of endemics, and epidemics and widespread health unhealthy conditions. In addition, the precariousness of the facilities and the dense occupation favored the occurrence of large-scale fires, increasing their impacts.

Figure 4:   Views of the Educandos district and the mouth of the Quarenta stream. *(Source: Google Maps.)*

We propose the renaturalization of the mouth Quarenta stream that involves in addition to the dimensions of infrastructure; architecture, and urbanism; public and socio-environmental governance also involves the historical and cultural dimension (Fig. 4). Which, from preliminary actions such as surveying the area and identifying problems and potential; families' registration that still live on stilts; the mapping the access to basic sanitation situation; identification of areas subject to relocation, and the land regularization that guarantees citizenship and social justice. In this sense, guidelines must be defined to overcome the following challenges:

i.   Civil Engineering Challenges: Execution of macro and micro-drainage, with the implementation of reservoirs and restoration of natural gutters; The Collection and Transport of Civil Construction Waste; Implementation of basic sanitation with innovative technological solutions for areas subject to flooding; The Depollution of the Quarenta/Educandos stream basin;

ii. Architecture and Urbanism Challenges: urbanization and requalification of houses with innovative, sustainable housing (a reinterpretation of the stilts), respecting the logic of the community's social organization, while restoring the meandering route of the stream;

iii. Public Governance Challenges: Land tenure regularization without major displacements; The relocation in an area of the industrial district of enterprises/ industries, which promote the development of micro and small businesses, common in the district (recyclable material collectors, locksmiths, furniture makers, etc.), organized in cooperatives and associations with training and incentives for adding values to its products, fostering the circular economy, which generates wealth for those workers. It also encourages the production of alternative materials for civil construction, such as ecological tile (produced from long-life packaging), panels and building blocks, produced from the reuse of materials; The regulation of the profession of recyclable material collectors, providing the means to carry out this work with dignity; The inclusion in urban planning of a rigorous environmental education policy; The promotion of tourism to the created historic site and the renaturalized hydrological basin;

iv. Social and environmental challenges: The development of ongoing environmental education programs; Involvement of residents and beneficiaries in processes to qualify local labor for ongoing actions, maintenance, and inspection focused on socio-environmental issues;

v. Historical and Cultural Dimension Challenges: The requalification of the stilts and creation of the historical site of the stilts, the floating city, and the landscape, with the watercourses being the structures of urban development – a historic repair

5 CONCLUSIONS

The successful experience of Mindu restoration encourages the struggle for the restoration of the Quarenta/Educandos stream, due to socio-environmental issues. It is also necessary the historical site creation of the stilts and the floating city, for the sake of social justice and cultural preservation.

The reinterpretation of the stilts and the creation of the floating city's is, above all, a matter of social justice for the native peoples, who since the rubber cycle have resisted the invisibility imposed by the public authorities. And it represents the reparation to the Amazonian historical and cultural heritage, for the important role of keeping the memory of its people and strengthening its roots.

The peculiarities of Manaus, such as the hydrographic basins magnitude that still have their springs preserved, the low population density despite being the seventh most populous city; one of the biggest tourist destinations in Brazil and, especially, being located in the center of the largest tropical forest in the world, it should lead its managers to consider the renaturalization of its watercourses, making them structural elements of urban infrastructure.

In addition, the watercourses renaturalization of Manaus, benefits the sustainability of the hydrographic basin, through the restoration of the drainage network and the streams renaturalization. It also has the potential to expand the perception of the Manaus population to issues related to sustainability and socio-environmental awareness. Moreover, the possibilities of boosting the economy and attracting investment in tourism. And in this way, contributing to a paradigm shift in public management.

The renaturalization of watercourses is essential for the rivers' preservation and the other water reserves. It necessarily depends on greater social and environmental responsibility, both on the part of politicians and the entire society involved.

ACKNOWLEDGEMENTS

To Dr. Myrian Koifman and Dr. Sybil V. Ferreira, civil engineers at SEMINF – Infrastructure Secretariat of Manaus/AM, both worked on the Mindu Restoration project, in all its phases, and are currently working on monitoring the actions. Gratitude for the encouragement and information provided regarding the Restoration of Mindu.

REFERENCES

[1] National Institute for Space Research (INPE), *Estudo do INPE indica que o rio Amazonas é 140 km mais extenso do que o Nilo,* Ministério da Ciência, Tecnologia e Inovações: São José dos Campos, 2008.

[2] National Water and Sanitation Agency (ANA), *Projeto Amazonas,* 8° Forum Mundial das Águas: Brazil: 2018.

[3] Criado, M.A., 500 Barragens que ameaçam o Rio Amazonas, *El País Journal*, 2017.

[4] Smith, N.J.H., *Amazon Sweet Sea: Land, Life, and Water at the River's Mouth,* University of Texas Press: Austin, TX, 2002.

[5] SGB-CPRM, Serviço Geológico do Brasil, *Rio Negro atinge 30 metros, na maior cheia em 119 anos em Manaus,* Ministério de Minas e Energia: Manaus, 2021.

[6] Riley. A.L., *Restoring Streams in Cities: A Guide for Planners, Policy Makers, and Citizens*, Island Press: Washington DC, 1998.

[7] Macedo. D.R., Callisto. M., Magalhães Jr., A.P., Restauração de Cursos d'água em Áreas Urbanizadas: Perspectivas para a Realidade Brasileira. *Revista Brasileira de REcursos Hídricos (RBRH),* **16**(3), pp. 127–139, 2011.

[8] Wade, P.M., Large, A.R.G. & de Wall, L.C., Rehabilitation of degrade rivers habitat. *Rehabilitation of Rivers Principles and Implementation*, John Wiley: Chichester, pp. 1–10, 1998.

[9] Bernhardt, E.S. & Palmer, M.A., Restoring strims in a urbanizing world. *Fresh Biology,* **52**(4), pp. 738–751, 2007. DOI: 10.1111/j.1365-2427.2006.01718.x.

[10] Lüderitz, V., Jüpner, R., Müller, S. & Felds, C.K., *Renaturalization of Streams and Rivers: The Special Importance of Integrated Ecological Methods in Measurement of Sucess. An Example from Saxony-Anhalt (Germany)*, Elsevier: Magdeburg, 2004.

[11] Wrong, T.H.F., Rogers, B.C. & Brown, R.R., *Transforming Cities through Water-Sensitive Principles and Pratices,* Elsevier: Melbourne, 2020.

[12] Travassos, L.R.F.C., *Revelando os Rios: Novos paradigmas para a intervenção em fundos de vale urbanos na Cidade de São Paulo,* Universidade de São Paulo: São Paulo, SP, 2010.

[13] Limeira, M.C.M., Silva, T.C. & Candido, G.A., Gestão adaptativa e sustentável para a restauração de Rios: Parte I – Enfoques teóricos sobre capacitação social. *Revista Brasileira de Recursos Hídricos,* **15**(1), pp. 17–26, 2010.

[14] Instituto de Proteção Ambiental do Amazonas (IPAAM), RIMA – Relatório de Impactos Ambientais para a Revitalização do Igarapé do Mindu, IPAAM: Manaus, AM, 2008.

[15] Serra, C.L.R. & Da Cruz, W.R., Aspectos econômicos e sociais da Cidade Flutuante, Gráfica Amazonas: Manaus, AM, 1964.

[16] Souza. L.J.B., *"Cidade Flutuante" uma Manaus sobre as águas (1920–1967),* Pontifícia Universidade Católica de São Paulo: São Paulo, SP, 2010.

PRIVATE LOT FLOOD PEAK ATTENUATION BY STORMWATER DETENTION TANKS

LUCY MARTA SCHELLIN[1], REBECCA MOREIRA DZIEDZIC[2] & MAURÍCIO DZIEDZIC[3]
[1]Secretaria Municipal de Meio Ambiente, Departamento de Recursos Hídricos, Brazil
[2]Department of Building, Civil and Environmental Engineering, Concordia University, Canada
[3]School of Engineering, University of Northern British Columbia, Canada

ABSTRACT

Changes in the hydrological cycle due to climate change and urbanization augment and accelerate runoff and flooding, degrade the urban environment, and cause human and material losses. Thus, it is important to implement measures that ensure urban hydrological conditions are kept as close as possible to pre-urbanization conditions, preventing floods. In addition to the conventional major and minor systems, cities may establish criteria for percentage of permeable area as well as stormwater management practices such as stormwater detention tanks, a type of low impact development technology (LID). The present study evaluates the adequacy of current practices in private lot detention tank design. It analyses time to empty, total detention time and flood peak abatement provided by detention tanks designed according to Curitiba's (Brazil) Bylaw 176/2007. Based on the results obtained, modifications were suggested to existing legislation to increase the efficiency of the detention tanks and, thus, reduce urban flooding and adapt to climate change. The proposed methodology can be applied elsewhere to guide detention tank design.
Keywords: flood control mechanisms, low impact development technologies (LID), sustainable drainage, sustainable hydrology, urban drainage.

1 INTRODUCTION

Urbanization drastically changes the hydrological characteristics of urban agglomerations. Balance is altered, water is polluted, and human populations increase, fostering urban growth to the detriment of local ecosystems [1]. In addition, hydrological changes associated with climate change have been observed and documented globally. These indicate that greater water storage can balance the water cycle and offset vulnerabilities related to flooding and water resource availability [2]. Increases in frequency and magnitude of flood events are expected due to climate change, with tropical and subtropical regions being more susceptible to flood events [3]. As a result, cost of water related services are expected to increase, as well as the cumulative risk of water shortages, floods and water quality degradation [4]. Accordingly, adaptation to climate change should consider measures of resilience [5], taking into account population spatial distribution and their access to urban infrastructure, public policies, among other factors [6]–[8].

To analyze the effects of intense rainfall and the performance of stormwater management facilities, hydrological models are used to estimate runoff distribution and intensity [8]–[10]. According to Hong [11], the importance of storing excess precipitation water in private flood holding reservoirs has been recognized in recent decades. These reservoirs can reduce the level or even eliminate the occurrence of floods. On-site stormwater detention (OSD) is a component of urban drainage systems [12] and a low impact development (LID) technology that helps reduce runoff and avoid overloading public stormwater infrastructure [13]. It can also contribute to the reduction of spatiotemporal variability in both local and river basin scales [2], and mitigate the effects of climate change in urban watersheds [14]. Furthermore, in water scarce contexts, these reservoirs can be used to harvest rainwater [15], [16].

WIT Transactions on The Built Environment, Vol 208, © 2022 WIT Press
www.witpress.com, ISSN 1743-3509 (on-line)
doi:10.2495/FRIAR220051

LID encompasses changes in the design and use of buildings and infrastructure to minimize impacts, being an important aspect of the sustainable management of stormwater [17], [18]. Considering the hydrological cycle, actions focus on individual lots and overall improvements are expected as a cumulative impact across the urbanized area. LID also seeks to keep impermeable surfaces to a minimum, and stormwater within the lot as long as possible [1].

The present study focuses on the application of stormwater tanks in the city of Curitiba, in southern Brazil. The city has a subtropical climate and most of its area has separate sewage and stormwater systems. Many low-income settlements are located in flood plains, making this part of the population particularly vulnerable to flooding. The city has been subjected to flooding since 1911, even before the consolidation of urban areas, and the situation has only worsened over the years. For this reason, legislation has existed since 1991 to mandate the construction of stormwater detention tanks on private properties to control runoff. These tanks decrease peak flows and complement macro drainage structures in the six main rivers that flow through the city. The legislation was revised in 2007, increasing the scope of buildings that can contribute to flood control in the municipality. This legislation establishes the minimum permeable area per lot, minimum detention tank volumes, and the diameter of the flow regulating orifice (FRO) but does not indicate the corresponding detention time, although it proposes a minimum of 20 minutes. Thus, a new revision is needed to promote ways to increase the efficiency of individual reservoirs on private lots to mitigate the effects of climate change. Limits for maximum specific outflow and detention times vary by municipality. Porto Alegre legislation [19] indicates a maximum specific outflow of 20.8 $L.s^{-1}.ha^{-1}$. In the city of São José do Rio Preto, the legislation [20] establishes a maximum outflow of 13 $L.s^{-1}.ha^{-1}$, while Curitiba's Drainage Master Plan [21] indicates 27 $L.s^{-1}.ha^{-1}$. No technical justifications are presented for these limits.

Most combinations of tank volumes and Flow Regulating Orifices (FRO) diameters established in Decree 176/2007 [22] are capable of holding peak floods for at least 20 minutes. Other municipalities recommend in their legislation detention times of one hour [23]–[26]. In other localities, systems that allow the infiltration of stormwater in the soil to reduce runoff and promote aquifer recharge are also prioritized [27]–[30].

The objective of the present study was to model peak flow reduction and detention times associated with prescribed tanks and FROs in order to propose changes to existing legislation and increase the efficiency of these devices.

2 METHODS

In order to model the efficiency of different tank sizes and FRO diameter combinations, tank emptying time and flood peak attenuation was calculated. The present methodology focuses on the tank volumes and FRO diameters established by Decree 176/2007 [22] and the local precipitation intensity but can be widely applied.

2.1 Tank emptying time

Unsteady flow analysis was performed to determine tank emptying times considering the volume ranges covered by the decree and above as they may be needed to promote lower peak discharges. Volumes analysed were between less than 2 m^3 and 20,000 m^3. Respective FRO diameters were based on commercially available PVC diameters, between 25 mm and 500 mm. Fig. 1 shows plan and section views of a typical detention tank. A vertical wall, the septum, separates the tank proper from an adjacent inspection chamber. These are connected at the bottom by the FRO, which is usually built with bricks or concrete blocks. Septum thickness generally varies between 12 and 19 cm, and an average 15.5 cm value is considered here. According to Finnemore and Franzini [31], if a wall has a thickness of less than 1.5 times the orifice diameter it can be considered thin, otherwise thick. Thus, for FRO diameters of up to 10 cm the septum is considered a thick wall, while above this FRO diameter the septum becomes a thin wall. This leads to a discharge coefficient, Cd, equal to 0.86 for tanks with FRO diameters equal to or less than 100 mm, and 0.61 otherwise [31].

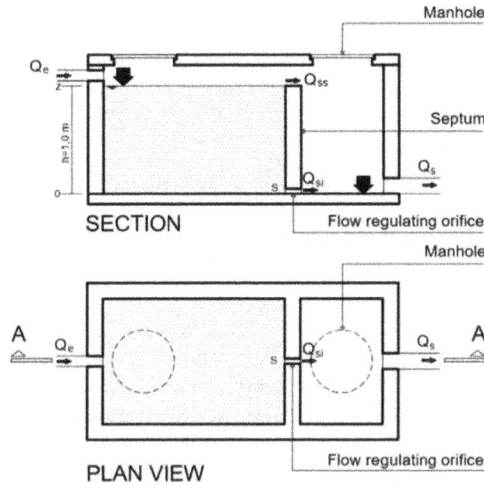

Figure 1: Plan and section views of a typical stormwater detention tank.

The discharge coefficient must be corrected for the position of the orifice (incomplete contraction weir) (eqn (1)). Here, a correction factor k = 0.5 was used since the orifice is usually installed at the bottom to allow full discharge, or in the middle or near one of the side tank walls [32], [33].

$$C_{d_{corrected}} = C_d \cdot (1 + 0.13.k), \tag{1}$$

where:
C_d = Discharge coefficient with complete contraction;
Correction factors for incomplete contraction:
$k = 0.25$ orifice near the bottom or near a wall;
$k = 0.5$ orifice near the bottom and near one wall;
$k = 0.5$ orifice near the bottom and two walls.

After this correction, the discharge coefficient becomes 0.92 for FRO diameters up to 100 mm and 0.65 otherwise.

Septum height is usually limited by the invert elevation of existing stormwater infrastructure, in order to guarantee gravity flow, with pumping as a last resort. Motor-pump assemblies are vulnerable to power outages as they can cause disturbances such as underground flooding in parking garages, impairment of drinking water tanks and fire prevention equipment, among others. Such assemblies also require frequent maintenance. Internal tank height (septum + overflow space) should be sufficient to allow at least one person for sediment removal, FRO clearing and tank maintenance. The recommendation for closed tanks is at least 80 cm of internal total height, the same minimum measure indicated in Decree 176/2007 for dry access chamber for maintenance [22]. Septum heights of 1, 1.2, 1.5, 1.8 and 2 m were considered, as these are the most often applied in existing projects and usually lead to feasible designs.

Emptying times were calculated considering a full tank with no inflow. These were obtained from eqn (2) (adapted from [31]).

$$\Delta t = \frac{A_r}{S} \cdot \left(\frac{1+C_d}{2g}\right)^{1/2} \cdot \frac{h^{1/2}}{1/2},$$ (2)

where:
Δt = tank emptying time (s);
A_r = tank surface area (m^2);
h = septum height or tank depth (m);
S = cross-sectional area of the flow regulating orifice (m^2);
g = acceleration of gravity 9.81 (m.s^{-2});
C_d = FRO discharge coefficient (0.92 and 0.65).

2.2 Flood peak attenuation

The analysis detention tank levels based on inflow applied eqns (3) to (8). The same variables represented in Fig. 1 were used. A septum height of 1.00 m was considered, since this is the most employed in projects, as it facilitates gravity flow to public drainage pipes, which, as a rule, are at a depth of 1.20 m.

Precipitation intensity is specific to each city and it his case was estimated using eqn (3) established by Parigot de Souza for the municipality of Curitiba [34], [35].

$$i = \frac{5950.T_R^{0,217}}{(t+26)^{1,15}},$$ (3)

where:
i = maximum rainfall intensity (mm.h^{-1});
T = return period (years);
t = rain duration time (min).

Return periods of 2 and 10 years were considered, usual for microdrainage works [36], [37]. A rain duration of 10 minutes was analysed, which is the magnitude of the time of concentration in the region analyzed [38].

The rational method was applied to estimate the outflow hydrograph, dependent upon peak lot outflow and time of concentration (T_c).

Considering that (T_c) generally depends on drainage basin factors and this study focuses on the operation of detention tanks, several T_c equations were tested. These were Hataway, Kirpich Tenessee, Kirpich Pennsylvania, FAA [39], NRCS (velocity), Kinematic wave [40], Temez, Bransby-Williams [41], Dooge, SCS (lag time), Izzard, Kirpich corrected [41], Ventura, Giandotti, Picking, Ven te Chow, George Ribeiro, Schaake et al. [42], Arizona DOT [44], FAA (2006), Papadakis-Kazan [45], Williams, Johnstone-Cross, Simas-Hawkins and Haktanir [46], UFCD [27], MPCA [28].

According to McCuen et al. [39], the non-conformity of a basin parameter for which the time of concentration equation was obtained is not a reason to discard the equation. A more comprehensive assessment of the study site is required. Results obtained by Silveira [47] showed that some equations obtained for rural basins showed good results in urban settings, even without the application of correction coefficients. Some equations are indicated for small drainage basins, but there is no consensus as to the maximum size that defines a small basin [27], [36], [42], [48]. Still, some of the small basin equations have also presented good results for large basins.

It should be noted, however, that T_c equations were not obtained for areas as small as individual urban lots, as is the case in the present work. Thus, no existing T_c equation is adequate, a priori, for the present case, and results need to be analyzed before being used. The time of concentration is used in urban drainage mainly for design. Overestimates lead to undersized drainage works, and vice versa. The present work focuses on urban lots, their drainage being the responsibility of the owners, and an attempt should be made not to penalize private property owners to pay for oversized drainage elements. Thus, the minimum value of 6.0 minutes recommended by the USDA [40] was adopted as a limit for the time of concentration.

Results below the minimum were discarded as well as those that were more than one standard deviation above the average. By applying the minimum concentration time, peak reduction results are more conservative. Lower concentration times would lead to steeper hydrographs and higher peak reduction would be attributed to the reservoirs. Eqn (4) was adopted [44] since it resulted in values closer to the average of the remaining results.

$$T_c = 3,258. \left(\frac{L_c}{S_c} \right)^{0,5} , \qquad (4)$$

where:
T_c = time of concentration (min);
L_c = talweg length (km);
S_c = slope (m.m^{-1}).

Tank peak inflow was calculated by eqn (5).

$$Q_e = C.i.A , \qquad (5)$$

where:
C = runoff coefficient (Rational Method);
i = precipitation intensity (eqn (2)) (mm.h^{-1});
A = catchment area (m^2).

Outflow through the FRO was given by eqn (6).

$$Q_{si} = S. C_{d\ corrected} \sqrt{2 \times 9.81 \times h}, \qquad (6)$$

where:
$C_{d\ corrected}$ = corrected discharge coefficient (eqn (1));
h = water level in the tank above the FRO (m).

Tank level variation (eqn (7)) over a time interval, Δt, was calculated by considering the difference between inflow and the outflow. When the bottom of the tank is permeable, the infiltration flow, Qi, must also be considered.

$$\Delta z = \frac{(Q_e - Q_{si} - Q_i).\Delta t}{A_r} . \qquad (7)$$

When the tank level exceeds the height of the septum overflow occurs, and eqn (8) is used.

$$Q_s = Q_{si} + Q_{ss} + Q_i, \tag{8}$$

where:
Q_s = outflow (m³.s⁻¹);
Q_{si} = outflow through the FRO (m³.s⁻¹);
Q_{ss} = overflow (over the septum) (m³.s⁻¹);
Q_i = infiltration in the soil (m³.s⁻¹).

Using eqns (3) to (8), an outflow hydrograph can be obtained. Post-urbanization hydrographs, with and without detention tanks were compared to the pre-urbanization situation, considering the lot covered with vegetation.

Eqn (9) is currently employed to calculate tank volume [22]:

$$V = k.i.A \tag{9}$$

where:
V = minimum tank volume (m³);
k = dimensionless constant = 0.2;
i = precipitation intensity = 80 mm.h⁻¹;
A = impervious area on the lot (m²).

The dimensionless constant k corresponds to the expected minimum detention time (1/3 of an hour, according to the current project practice in the municipality) multiplied by the difference between the impermeability coefficients adopted: current (0.9) and pre-urbanization (0.3).

3 RESULTS AND DISCUSSION

A typical hydrograph obtained through the procedures summarized by eqns (7) and (8) is shown in Fig. 2. It shows a lot in pre-urbanization and post-urbanization conditions, with and without detention tanks. The dashed line represents the hydrograph before urbanization with 75% grass cover, 0 to 2% slope, 10 years return period precipitation, soil type C and runoff coefficient of 0.25 [42]. The dotted line represents the surface runoff for concrete and roof surfaces, 10 years return period precipitation, type C soil, and runoff coefficient of 0.83) [42]. The continuous line represents the surface runoff with the implementation of a detention tank [22].

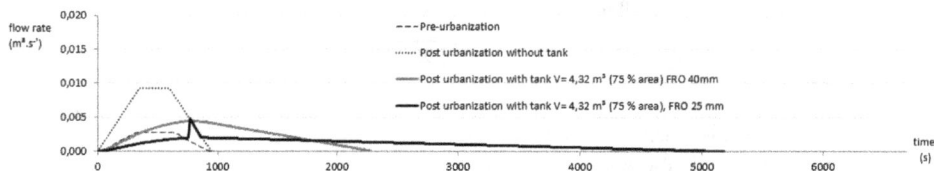

Figure 2:   Hydrograph for precipitation with T_2 and t_{10}, FRO of 40 and 25 mm, and tank volume calculated on 75% of area.

The precipitation intensity of 80 mm.h⁻¹ considered in Decree 176/2007 [22] is below the T_2 to T_{10} values recommended by standards for urban drainage [36], [37]. The present study compared scenarios with t_2, t_{10} and T_{10}, calculated according to eqn (3). With T_2, and t_{10}, the precipitation intensity results in $i_{2,10} = 112.23$ mm.h⁻¹, while T_{10} gives $i_{10,10} = 159.14$ mm.h⁻¹. An increase of 20% in rainfall intensity was considered, as suggested by the

Table 1:   Total detention times and average outflow as a function of tank volume and flow regulating orifice (FRO) diameters recommended here.

| FRO diameter (mm) | Tank volume (m³) | Septum height (m). | | | | | | | | | |
| | | h = 1.00 | | h = 1.20 | | h = 1.50 | | h = 1.80 | | h = 2.00 | |
		Time (min.)	Average outflow (L.s⁻¹)	Time (min.)	Average outflow (L.s⁻¹)	Time (min.)	Average outflow (L.s⁻¹)	Time (min.)	Average outflow (L.s⁻¹)	Time (min.)	Average outflow (L.s⁻¹)
25	≤ 3.0	≤ 64	0.79	≤ 58	0.86	≤ 52	0.96	≤ 47	1.05	≤ 45	1.11
40	3.1 to 15.9	26 to 132	2.01	23 to 129	2.20	21 to 108	2.46	19 to 98	2.70	18 to 93	2.84
40 C_d 0.92	7.24	60	2.01	55	2.20	49	2.46	45	2.70	42	2.84
40	10.24	85	2.01	77	2.20	69	2.46	63	2.70	60	2.84
50	16.0 to 35.9	85 to 190	3.14	77 to 174	3.44	69 to 155	3.85	63 to 142	4.22	60 to 135	4.44
75	36.0 to 63.9	85 to 151	7.07	77 to 138	7.74	69 to 123	8.66	63 to 112	9.48	60 to 107	10.00
100	64.0 to 155.9	86 to 207	12.57	78 to 189	13.77	69 to 169	15.39	63 to 154	16.86	60 to 146	17.77
150	156.0 to 275.9	85 to 151	30.47	78 to 138	33.38	70 to 123	37.32	64 to 112	40.88	60 to 107	43.09
200 C_d 0.65	276.0 to 620.9	85 to 191	54.17	78 to 174	59.34	69 to 156	66.35	63 to 142	72.68	60 to 135	76.61
300	621.0 to 1,104.9	85 to 151	121.89	78 to 138	133.52	69 to 123	149.28	63 to 113	163.53	60 to 107	172.37
400	1,005.0 to 1,725.9	85 to 133	216.69	78 to 121	237.37	69 to 108	265.39	63 to 99	290.72	60 to 94	306.44
500	≥ 1,726	≥ 85	338.57	≥ 78	370.89	≥ 69	414.67	≥ 63	454.25	≥ 60	478.82

Intergovernmental Panel on Climate Change [49] for the Curitiba region (Magrin et al., 2007; Marengo, 2008), resulting $i_{2,10}$ + 20% = 134.67 mm.h^{-1} and $i_{10,10}$ + 20% = 190.96 mm.h^{-1}.

For precipitation intensity 112.23 mm.h^{-1}, increasing the reservoir volume increases the detention time by 12%, while changing the FROs more than doubles the detention time, even with some overflow (FRO 25 mm). Thus, combining reservoir volumes with the appropriate FROs leads to better results than simply increasing reservoir volumes. The higher the volume of the tank, the lower the probability of septum overflow. Peak flow control is most efficient when there is no overflow. However, by simply increasing FRO diameter to avoid septum overflow, detention time is reduced, and rainfall is released to the receiving water body sooner, which can reduce flood control efficiency. Thus, considering the inherent septum height limitations, the best condition is achieved by adequate combination between tank volume and FRO diameter, in order to optimize flood peak abatement.

Table 1 presents combinations of tank volumes, FRO diameters, and septum heights recommended here. A minimum emptying time of one hour is achieved for all combinations, except for a 25 mm FRO combined with septa higher than 1.00 m. FRO diameters below 25 mm can be easily clogged and compromise tank operation, since water could remain inside the tank, lacking space to accommodate the next rainfall. This would also lead to high detention times, conducive to the development of aquatic organisms such as mosquito larvae. Thus, 25 mm FROs require more frequent maintenance, and are not generally recommended.

Table 2 shows detention tank volumes and the corresponding FRO diameters in accordance with Decree 176/2007 and the proposed changes. These tank size – FRO diameter combinations ensure flood peak reductions for all cases studied.

Table 2: Suggested tank volume and FRO diameter combinations.

Tank size (m^3) Decree 176/2007	Tank size (m^3) Proposed	FRO (mm)
≤ 2	≤ 3	25
3 to 6	3.1 to 15.9	40
7 to 26	16.0 to 35.9	50
27 to 60	36.0 to 64.9	75
61 to 134	65.0 to 155.9	100
135 to 355	156.0 to 275.9	150
356 to 405	276.0 to 620.9	200
406 to 800	621.0 to 1.104.9	300
801 to 1,300	1,105.0 to 1,725.9	400
1,301 to 2,000	≥ 1,726.0	500

4 CONCLUSIONS

Notwithstanding the fact that Decree 176/2007 was supposedly conceived considering a rainfall intensity of 80 mm.h^{-1} and a 20 minute detention time, results obtained here show that tank volume and FRO diameter combinations prescribed by this decree are effective for higher rainfall intensities and produce longer detention times. The only exception is the tank volume range for 25 mm FROs.

Stormwater detention tanks smaller than 3.0 m^3 are employed in small lots, usually in low-income housing developments [50]. The combined effect of several such tanks is not negligible [35], but they require careful maintenance, as 25 mm FROs are easily obstructed. Thus, the use of other LID technologies such as permeable pavement may increase flood control efficiency. Achieving the total detention design volume through the combination of smaller tanks is another valid option. It might facilitate runoff collection due to lot topography and reduced costs. Partitioning the volume may also help to avoid very long detention times, which could lead to the proliferation of disease vectors. The results presented herein indicate that stormwater detention tanks can be used to reduce the effects of urbanization in the hydrological cycle, since they abate flood peaks and also lengthen the base of the hydrograph. However, stormwater detention tanks, even if they promote the abatement of flood peaks, cannot be considered the only solution to control runoff. It is possible, and necessary, to combine their use with that of other LID technologies that promote evapotranspiration and infiltration of stormwater into the ground to recharge aquifers. Green corridors (linear parks), containment basins, rooftop detention systems (green roofs, blue roofs), and others, are among such technologies that, in addition to flood control, contribute to pollutant removal and water quality improvement in urban water bodies.

The method proposed and demonstrated herein to adjust flood control legislation in the city of Curitiba can be used to perform similar analysis elsewhere. In such cases, rainfall intensities would assume different values, and local soil conditions would have to be considered to account for different infiltration rates. With those changes, the overall method remains the same, and unsteady flow analysis is employed to determine detention times and the corresponding hydrographs.

REFERENCES

[1] Davis, A.P., Green engineering principles promote low-impact development. *Environmental Science and Technology*, **39**(16), 2005. DOI: 10.1021/es053327e.

[2] Ehsani, N., Vörösmarty, C.J., Fekete, B.M. & Stakhiv, E.Z., Reservoir operations under climate change: Storage capacity options to mitigate risk. *Journal of Hydrology*, **555**, pp. 435–446, 2017. DOI: 10.1016/j.jhydrol.2017.09.008.

[3] Eccles, R., Zhang, H. & Hamilton, D., A review of the effects of climate change on riverine flooding in subtropical and tropical regions. *Journal of Water and Climate Change*, **10**(4), pp. 687–707, 2019. DOI: 10.2166/wcc.2019.175.

[4] Coombes, P.J., Status of transforming stormwater drainage to a systems approach to urban water cycle management: Moving beyond green pilots. *Australian Journal of Water Resources*, **22**(1), pp. 15–28, 2018. DOI: 10.1080/13241583.2018.1465376.

[5] Charlesworth, S.M., A review of the adaptation and mitigation of global climate change using sustainable drainage in cities. *Journal of Water and Climate Change*, **1**(3), pp. 165–180, 2010. DOI: 10.2166/wcc.2010.035.

[6] Choi, H., Lee, E.H., Joo, J.G. & Kim, J.H., Determining optimal locations for rainwater storage sites with the goal of reducing urban inundation damage costs. *KSCE Journal of Civil Engineering*, **21**(6), pp. 2488–2500, 2017. DOI: 10.1007/s12205-016-0922-6.

[7] Chow, W.T.L., The impact of weather extremes on urban resilience to hydro-climate hazards: A Singapore case study. *International Journal of Water Resources Development*, **34**(4), pp. 510–524, 2018. DOI: 10.1080/07900627.2017.1335186.

[8] Moore, T.L., Gulliver, J.S., Stack, L. & Simpson, M.H., Stormwater management and climate change: Vulnerability and capacity for adaptation in urban and suburban contexts. *Climatic Change*, **138**(3–4), pp. 491–504, 2016. DOI: 10.1007/s10584-016-1766-2.

[9] Guo, Y. & Zhuge, Z., Analytical probabilistic flood routing for urban stormwater management purposes. *Canadian Journal of Civil Engineering*, **35**(5), pp. 487–499, 2008. DOI: 10.1139/L07-131.

[10] Peng, J., Ouyang, J. & Yu, L., The model of low impact development of a sponge airport: A case study of Beijing Daxing international airport. *Journal of Water and Climate Change*, **12**(1), pp. 116–126, 2021. DOI: 10.2166/wcc.2020.195.

[11] Hong, Y.M., Experimental evaluation of design methods for in-site detention ponds. *International Journal of Sediment Research*, **25**(1), pp. 52–63, 2010. DOI: 10.1016/S1001-6279(10)60027-3.

[12] Ronalds, R., Rowlands, A. & Zhang, H., On-site stormwater detention for Australian development projects: Does it meet frequent flow management objectives? *Water Science and Engineering*, **12**(1), pp. 1–10, 2019. DOI: 10.1016/j.wse.2019.03.004.

[13] Chang, N.-B., Hydrological connections between low-impact development, watershed best management practices, and sustainable development. *Journal of Hydrologic Engineering*, **15**(6), pp. 384–385, 2010. DOI: 10.1061/(asce)he.1943-5584.0000236.

[14] Andrés-Doménech, I., Montanari, A. & Marco, J.B., Efficiency of storm detention tanks for urban drainage systems under climate variability. *Journal of Water Resources Planning and Management*, **138**(1), pp. 36–46, 2012. DOI: 10.1061/(asce)wr.1943-5452.0000144.

[15] Shadmehri Toosi, A., Danesh, S., Ghasemi Tousi, E. & Doulabian, S., Annual and seasonal reliability of urban rainwater harvesting system under climate change. *Sustainable Cities and Society*, **63**(May), 102427, 2020. DOI: 10.1016/j.scs.2020.102427.

[16] Sheikh, V., Perception of domestic rainwater harvesting by Iranian citizens. *Sustainable Cities and Society*, **60**(May), 102278, 2020. DOI: 10.1016/j.scs.2020.102278.

[17] Gimenez-Maranges, M., Breuste, J. & Hof, A., A new analytical tool for a more deliberate implementation of sustainable drainage systems. *Sustainable Cities and Society*, **71**, 2021. DOI: 10.1016/j.scs.2021.102955.

[18] Gimenez-Maranges, M., Pappalardo, V., La Rosa, D., Breuste, J. & Hof, A., The transition to adaptive storm-water management: Learning from existing experiences in Italy and southern France. *Sustainable Cities and Society*, **55**(January), 102061, 2020. DOI: 10.1016/j.scs.2020.102061.

[19] Decree 18611, (testimony of Porto_Alegre), 2014. http://www2.portoalegre.rs.gov.br/cgi-bin/nph-brs?s1=000033997.DOCN.&l=20&u=%2Fnetahtml%2Fsirel%2Fsimples.html&p=1&r=1&f=G&d=atos&SECT1=TEXT.

[20] Lei Ordinária 10290, (testimony of São José do Rio Preto), 2008. https://leismunicipais.com.br/a/sp/s/sao-jose-do-rio-preto/lei-ordinaria/2008/1029/10290/lci-ordinaria-n-10290-2008-cria-no-municipio-o-programa-permanente-de-gestao-das-aguas-superficiais-pgas-da-bacia-hidrografica-do-rio-preto-e-da-outras-providencias.

[21] CH2MHill, Plano de Drenagem do Alto Iguaçu. 2002. http://www.iat.pr.gov.br/Pagina/Plano-de-Drenagem-do-Alto-Iguacu.

[22] Decreto 176, 5 (testimony of Curitiba), 2007.

[23] Lei Ordinária 10290, (testimony of São José do Rio Preto), 2008.

[24] Lei Complementar 428, (testimony of São José dos Campos), 2010. https://www.sjc.sp.gov.br/legislacao/Leis Complementares/2010/428.pdf.

[25] Lei Ordinária 13276, (testimony of São Paulo), 2002. https://leismunicipais.com.br/a/sp/s/sao-paulo/lei-ordinaria/2002/1327/13276/lei-ordinaria-n-13276-2002-torna-obrigatoria-a-execucao-de-reservatorio-para-as-aguas-coletadas-por-coberturas-e-pavimentos-nos-lotes-edificados-ou-nao-que-tenham-area-impermeabilizada-superior-a-500m.

[26] Lei Ordinária 9952, (testimony of Sorocaba), 2012. https://leismunicipais.com.br/a1/
      sp/s/sorocaba/lei-ordinaria/2012/996/9952/lei-ordinaria-n-9952-2012-dispoe-sobre-
      normas-para-a-contencao-de-enchentes-e-destinacao-de-aguas-pluviais-e-da-outras-
      providencias?q=9952.
[27] Colorado, *Urban Storm Drainage Criteria Manual*, vol. 1, issue March, 2017.
      http://udfcd.org/criteria-manual.
[28] Minnesota, *The Minnesota Stormwater Manual*, 2005. https://www.leg.mn.gov/docs/
      2006/other/060302.pdf.
[29] NACTO, *Urban Street Stormwater Guide*, 2016. https://nacto.org/publication/urban-
      street-stormwater-guide/.
[30] Philadelphia Water, *Philadelphia Water Stormwater Plan Review*, 2020.
      https://www.pwdplanreview.org/manual-info/guidance-manual.
[31] Finnemore, E. & Franzini, J., *Fluid Mechanics with Engineering Applications*, 10th
      ed., McGraw-Hill Education, 2001.
[32] Horton, R.E., Weir experiments, coefficients, and formulas. Department of the
      Interior, United State Geological Survey, **200**, 1907.
[33] Netto, A. & Fernández, M.F.Y., *Manual de Hidráulica*, 9th ed., 2015.
[34] Pinto, N.L. de S., Holtz, A.C.T., Martins, J.A. & Gomide, F.L.S., *Hidrologia Básica*,
      14th ed., Blucher, 2013.
[35] Schellin, L.M. & Dziedzic, M., Comparação de equações de intensidade de chuva
      obtidas para Curitiba para cálculo de elementos de microdrenagem. *Anais XXII
      Simpósio Brasileiro de Recursos Hídricos 2*, ed. ABRH, pp. 1–7, 2017.
      https://anais.abrhidro.org.br/job.php?Job=2978.
[36] ASCETESB, Drenagem urbana: manual de projetos da CETESB/ASCETESB, 1986.
[37] São_Paulo, Manual de drenagem e manejo de águas pluviais, 2012.
[38] Lopes, E., Ri, C.A. & Gomes, J., Avaliação do impacto da redução do escoamento
      superficial sobre a vazão de saída em uma rede de microdrenagem. *Anais XIX SImpósio
      Brasileiro de Recursos Hídricos*, 2011. https://abrh.s3.sa-east-1.amazonaws.com/
      Sumarios/81/9c0fb98ac485a8452068adeb5f5719a1_ced6ae3fcd9bdf2fd522bde910c3
      0814.pdf.
[39] McCuen, R.H., Wong, S.L. & Rawls, W.J., Estimating urban time of concentration.
      Journal of Hydraulic Engineering, **110**(7), pp. 887–904, 1984.
      DOI: 10.1061/(asce)0733-9429(1984)110:7(887).
[40] USDA, Urban hydrology for small watersheds, 1986. www.nrcs.usda.gov/Internet/
      FSE_DOCUMENTS/stelprdb1044171.pdf.
[41] Salimi, E.T., Nohegar, A., Malekian, A., Hoseini, M. & Holisaz, A., Estimating time
      of concentration in large watersheds. *Paddy and Water Environment*, **15**(1), pp. 123–
      132, 2017. DOI: 10.1007/s10333-016-0534-2.
[42] Chow, V.T., Maidment, D.R. & Mays, L.W., Applied hydrology. *Applied Hydrology*,
      McGraw-Hill, 1988.
[43] FAA, Advisory circular: UFC 3-230-01, AC 150/5320-5C, 2006. https://www.faa.gov/
      documentLibrary/media/advisory_circular/150-5320-5C/150_5320_5c_p2.pdf.
[44] Fang, X., Thompson, D.B., Cleveland, T.G., Pradhan, P. & Malla, R., Time of
      concentration estimated using watershed parameters determined by automated and
      manual methods. *Journal of Irrigation and Drainage Engineering*, **134**(2), pp. 202–
      211, 2008. DOI: 10.1061/(asce)0733-9437(2008)134:2(202).
[45] USDA, *National Engineering Handbook*, 2008. http://irrigationtoolbox.com/NEH/
      Part630_Hydrology/NEH630-ch15draft.pdf.

[46] Arizona, Highway drainage design manual: Hydrology, 1994. https://apps.azdot.gov/files/ADOTLibrary/publications/project_reports/pdf/az442.pdf.

[47] Silveira, A., Desempenho de Fórmulas de Tempo de Concentração em Bacias Urbanas e Rurais. *Revista Brasileira de Recursos Hídricos*, **10**(1), pp. 5–29, 2005. DOI: 10.21168/rbrh.v10n1.p5-29.

[48] Guo, J. & Urbonas, B., Volume-based runoff coefficients for urban catchments. *Journal of Irrigation and Drainage Engineering*, **140**(2), 04013013, 2014. DOI: 10.1061/(asce)ir.1943-4774.0000674.

[49] IPCC, Impacts, adaptation and vulnerability. contribution of working group ii to the third assessment report of the Intergovernmental Panel on Climate Change, Cambridge University Press, 2014.

[50] Lei de Zoneamento, (testimony of Curitiba), 2000. https://www.curitiba.pr.gov.br/conteudo/nova-lei-de-zoneamento/141.

SECTION 3
MODELLING AND
EXPERIMENTATION

IDENTIFICATION OF SEWAGE EXFILTRATION IN COASTAL AREAS THROUGH THE MONITORING OF DRUGS AND STIMULANT CONCENTRATIONS IN URBAN STORM DRAINS

STEVEN SPENGLER[1] & MARVIN HESKETT[2]
[1]Pacific Rim Water Resources, Hawaiʻi
[2]Element Environmental, Hawaiʻi

ABSTRACT
One of the major barriers for municipalities responsible for mitigation of sewage exfiltration is locating grossly leaking sections of the sewage conveyance system in a time-, labor- and cost-efficient manner. In this study, water samples were collected from the dense network of manholes overlying the storm drain systems in the tourist area of Waikīkī and inland residential areas on the island of Oʻahu, Hawaiʻi. The majority of the sewage conveyance infrastructure in this coastal area is submerged and the storm drains are routinely subject to backflow during high tide. Exfiltration of sewage from the aging conveyance system in this coastal area contaminates the surrounding shallow brackish aquifer, which then enters leaking pipe joints and cracks in the storm water conveyance system. Samples collected from the storm drains were analyzed for the presence of carbamazepine, a commonly prescribed anticonvulsant, pain relief and bipolar disorder treatment drug, which behaves as a conservative tracer in the environment (> 50 days half-life, low sorption). Samples were also analyzed for the more labile anthropogenic tracer caffeine (~4 day half-life). The higher stability of carbamazepine enables detection of this compound at greater distances from sewage release sites while caffeine serves as a better tracer for detecting recent, proximal releases of sewage, given its ephemeral nature and relatively high and ubiquitous presence. The concentration levels and spatial distribution of detection of these two anthropogenic biomarkers were successfully used to identify areas of ongoing sewage exfiltration in Waikīkī and surrounding residential communities. The variation in carbamazepine and caffeine concentrations measured in Waikīkī storm drains over a 1 year period generally correlate with daily visitor arrivals to Oʻahu.
Keywords: sewage exfiltration, Hawaii, Waikiki, pharmaceutical tracer, storm drain contamination.

1 INTRODUCTION
An average of 394 million liters per day (mld) of sewage is conveyed through Oʻahu's 3,380 km web of underground sewer lines. The majority of sewer lines in urban Honolulu are over 65 years old with an overall age range distribution on Oʻahu as follows: <25 years; 22.2%, 26–50 years; 18.3%; 51–75 years; 42.7%, 76–100 years; 11.8% and >100 years, 1.2%. In coastal areas on Oʻahu, the sewage conveyance system is largely immersed in the shallow, brackish to saline groundwater aquifer that underlies the coastal plain. The warm climate in Hawaiʻi combined with the high sulfate content of the saline groundwater produces corrosive hydrogen sulphide gases that results in constant challenges in maintenance of the aging sewage conveyance system. Thus, there is an ongoing need to replace and upgrade Hawaiʻi's sewer lines and force mains due to both capacity and structural integrity issues [1].

The exfiltration of sewage in coastal settings such as Waikīkī contaminate the surrounding shallow brackish aquifer, which then enters leaking pipe joints and cracks in the portions of the aging storm water conveyance system that conveys tidal water inland from either the Ala Wai Canal or the Pacific Ocean during the upper part of the tidal cycle. One study by the United States Environmental Protection Agency reported between 12% and 49% of wastewater flows are lost due to leaking infrastructure in United States cities [2]. A recent

WIT Transactions on The Built Environment, Vol 208, © 2022 WIT Press
www.witpress.com, ISSN 1743-3509 (on-line)
doi:10.2495/FRIAR220061

survey completed by the Water Environment Foundation in Milwaukee (315 outfalls, n = 1,500 samples) estimated that 30% of stormwater outfalls show high and consistent levels of untreated wastewater and 8% had very high levels of wastewater.

The system-wide amount of wastewater exfiltration from Oʻahu's sewage system is unknown. A rough, upper-limit estimate of the magnitude of exfiltration can be made by comparing the average daily groundwater withdrawals on Oʻahu by the island-wide water utility, the Honolulu Board of Water Supply (BWS), during the wet winter months of January to March 2021 (127.6, 124.8 and 121 mgd, respectively) with the daily average volume of wastewater treated at the nine Waste Water Treatment Plants (WWTP) operated by the City and County of Honolulu (CCH) on Oʻahu during 2021 (103.7 mgd). The difference in the volume of groundwater pumped by the BWS during these wet months (when use of water for irrigation purposes is at a minimum) and the volume of wastewater is around 20 mgd. It is likely that at least half (10 mgd) of this difference is used during these wet winter months for purposes (residential irrigation, small-capacity private WWTP, wash cars, fill pools, etc.) that doesn't result in a return of the spent water to the sewer system. By comparison, the average daily groundwater withdrawal by the BWS during the dry summer months of June, July and August 2021 were 150, 151 and 150 mgd, a difference of around 45 mgd from the volume of water processed at the nine CCH WWTPs. This simple analysis suggests that the system-wide exfiltration rate on Oʻahu is somewhere in the order of 10%, with the majority of exfiltration likely occurring in the older sections of pipes present in urban Honolulu.

One of the major barriers for municipalities responsible for mitigation of sewage exfiltration is locating grossly leaking sections of the sewage conveyance system in a time-, labor- and cost-efficient manner. The City and County of Honolulu sewers are currently monitored and rehabilitated through extensive CCTV inspection of sewer lines, and a semi-automated computer algorithm to evaluate the CCTV results [1].

The primary objective of the study was to determine whether mapping of the spatial distribution of the maximum pharmaceutical (carbamazepine and caffeine) concentration levels measured in the storm drain systems in the vicinity of Waikīkī can be used to identify areas of on-going sewage exfiltration from the aging sewer conveyance systems in these areas. Ideally, this type of study would utilize shallow groundwater data collected from a dense network of monitoring wells within the study area. Unfortunately, no such monitoring network exists. So, in this study, samples were collected between November 2020 and January 2021 from the dense network of storm water manholes present in this coastal tourist destination and the adjacent, inland McCully-Moiliili residential areas. In these areas, wastewater contaminated groundwater seeps into the stormwater conveyance pipes and mixes with the tidally driven water that enters the storm drain system from the Ala Wai Canal or the ocean (dependent on which side of Waikīkī is sampled). A secondary objective of the study was to measure the temporal variations in caffeine and carbamazepine concentrations measured in three Waikīkī storm drains and at two Ala Wai Canal locations between February 2021 and December 2021, when the number of tourists visiting Hawaiʻi varied greatly due to COVID pandemic travel restrictions to the islands.

1.1 Caffeine and carbamazepine

Carbamazepine and caffeine are known as emerging contaminants, which describes pollutants that have been detected in water bodies, may have ecological or human impact, and typically are not regulated under current environmental laws. These compounds are also known as micropollutants because they are typically present in trace quantities (part per trillion to billion levels) in the environment. These micropollutants enter the environment during our daily routines when we consume, flush away, or wash these compounds down the sink. As a result, these compounds are increasingly being used as anthropogenic (human) markers of sewage contamination. Carbamazepine and caffeine were the first and second (62.3% and 56.1%, respectively) most commonly active pharmaceutical ingredients detected in 1,052 river samples collected from 104 countries worldwide in a recently published global-scale study [3]. Caffeine was detected in river samples collected from every continent while carbamazepine was detected in rivers on all continents except Antarctica. The detection frequency of caffeine and carbamazepine in rivers were similar across all six continents where both pharmaceutical compounds were detected.

Caffeine is a naturally occurring stimulant found in coffee, soda, tea, chocolate and energy drinks. The daily consumption rate of caffeine varies worldwide. Northern European countries tend to consume higher daily doses of caffeine (190 to 260 mg/day/person) compared to warmer Southern European countries (80 to 120 mg/day/person). The average daily consumption of caffeine in the United States is 165 mg per person per day [4]. Caffeine is extensively metabolized by humans during consumption, with less than 5% excreted unchanged in the urine [5].

Carbamazepine is a commonly prescribed anticonvulsant, pain relief and bipolar disorder treatment drug. The dosage range for carbamazepine for those on the medication is between 400 to 1,200 mg/day. Daily per capita consumption rates of carbamazepine range from 0.03 to 0.44 mg/day/person [6]. Following consumption, up to 10% of CBZ is excreted from the human body [7].

Caffeine is effectively removed (> 95%) by conventional treatment at WWTPs while carbamazepine is poorly removed (typically less than 10%). Caffeine was found to undergo significant microbial degradation in the Jamaica Bay estuary near New York City while little evidence for removal of carbamazepine was observed [8]. These findings suggest that carbamazepine behaves as a conservative tracer in the environment whereas caffeine is comparatively labile. The persistence of carbamazepine in conventional treatment processes leads to its widespread occurrence in water bodies, especially on the mainland United States where WWTPs commonly discharge treated effluent into nearby surface water bodies [9]. The stability of carbamazepine upon release to the environment enables detection at larger distances from sewage release sites while caffeine serves as a better tracer for detecting recent, proximal releases of sewage or grey-water, given its ephemeral nature and higher initial concentration levels in untreated wastewater.

1.2 Caffeine and carbamazepine concentration levels in sewage in Hawaiʻi

Researchers at the University of Hawaiʻi monitored the variation in sewage flow and caffeine concentration in the main sewer line exiting Manoa Valley on Oʻahu by collecting composite samples over 3 hour periods during two, week-long dry-weather monitoring periods [10]. They found that the concentration of caffeine showed reproducible daily patterns with the highest concentrations being observed in the mornings and at end of the day (i.e., 8–11 AM and 5–8 PM composite samples), which also corresponded to periods of generally higher sewage flow exiting Manoa Valley. The measured concentration of caffeine in the sewage ranged from 5,000 to 103,400 parts per trillion (ppt, ng/L), with the highest flux of caffeine (2.4 mg/sec) exiting the valley between 8 and 11 AM. The authors associated the primary source of caffeine with the preparation and consumption of coffee and tea, which is secreted to the sewer system by the regular daily metabolic activities of the residents in the valley.

The Safe Drinking Water Branch of the State of Hawaiʻi Department of Health (HDOH) conducted two rounds of sampling of raw wastewater influent at four WWTP and also collected thirteen samples of treated wastewater effluent generated at thirteen WWTP facilities that produce reclaimed water throughout the State of Hawai'i [11]. The HDOH samples were analyzed using LC-MS-MS methods. Carbamazepine was not detected in the wastewater influent; likely due it being masked by the chromatographic peaks of higher concentration analytes. During the current study, a total of twenty samples were collected from septic tanks at various beach parks on the island of O'ahu and analyzed for caffeine and carbamazepine using an ELISA immunoassay method. Table 1 summarizes the caffeine and carbamazepine concentrations measured. The median concentration of caffeine and carbamazepine is around 100,000 ppt and 500 ppt, respectively.

Table 1:   Caffeine and carbamazepine concentrations measured in septic tanks and WWTP effluent and influent in Hawai'i.

Source	Caffeine (parts per trillion)				
	Mean	Median	Max. Detect	Count	Detect Frequency
Septic Tanks	19,925	11,600	83,200	11	100%
WWTP Influent[1]	96,238	108,000	150,000	8	100%
WWTP Effluent[1]	152	33	1,200	23	96%
Source	Carbamazepine (parts per trillion)				
	Mean	Median	Max. Detect	Count	Detect Frequency
Septic Tanks	682	552	1,735	20	100%
WWTP Influent[1]	ND	ND	ND	8	0%
WWTP Effluent[1]	113	110	220	23	65%
ND = Not Detected					
[1]Oahu WWTP Influent/Efffluent Data from HDOH [11]					

2 DESCRIPTION OF STUDY AREA

This study was conducted on the island of O'ahu in the 4 km^2 tourist center of Waikīkī and the coastal portion of the adjacent residential communities of McCully-Moiliili, which are comprised of a mix of high rises, low rises and single-family residential homes. Roughly 75% of the State of Hawai'i's population of 1.44 million people reside on O'ahu. Waikīkī means "spouting water" in the Hawaiian language in reference to the rivers and springs that flowed into the area during historic times. The original marsh and swamp lands were considered a health hazard and drained during construction of the Ala Wai Canal in the 1920s, which directs streamflow and storm runoff from the Manoa and Palolo watersheds to the Pacific Ocean and separates Waikīkī from the inland McCully-Moiliili communities. During development of the McCully-Moiliili areas in the 1940s, inland stream flow, spring flow and storm runoff through these areas were routed into lined channels known as the Hausten and Makiki Ditches, which empty into the Ala Wai Canal.

Waikīkī became a major tourist destination in the 1950s with the advent of long-distance air travel to the islands from the mainland United States which prompted the beginning of the construction of the numerous high-rises and resort hotels that dominate today's skyline. As a result, most of the sewer and storm-drain infrastructure in the Waikīkī area is between 50 and 70 years old. Visitor arrivals reached one million for the first time in 1968, filling approximately 15,000 hotel rooms. From 1990 to 2013, between 4 and 5 million visitors arrived annually on the island of Oʻahu [12], and the number of hotel rooms in Waikīkī had expanded to 30,000 rooms. Between 2013 to 2019, annual tourist arrivals to Oʻahu increased to between 5 and 6 million visitors. Peak visitor arrivals tend to occur during the months of December and July and international visitors (largely from Japan) contribute around one-third of the travelers to Hawaiʻi. Travel restrictions imposed due to the COVID-19 pandemic caused the annual visitor arrivals to Oʻahu to plummet to 1.5 million in 2020 followed by a partial rebound of 3.3 million annual visitor arrivals in 2021.

The sampling effort presented in this study was conducted between November 2020 and December 2021, roughly 6 to 20 months into the global pandemic. The spatial distribution of sewage exfiltration in the Waikīkī and inland McCully-Moiliili areas was based on sampling conducted between November 2020 to January 2021 while temporal variations in caffeine and carbamazepine concentrations were monitored in three Waikīkī storm drain and two Ala Wai Canal locations between February 2021 and December 2021.

3 STUDY METHODOLOGY

A total of 70 storm drain and canal sampling locations were established within Waikīkī, the inland residential McCully-Moiliili communities and along the Ala Wai Canal which separates these two areas. The Ala Wai Canal receives input from the storm drain systems that underlie both the Waikīkī and the inland residential McCully-Moiliili communities. Samples were collected directly from the storm drains by inserting a 1.3 cm diameter PVC pipe through the vent hole in the storm drain manhole cover to the bottom of the storm drain, capping the top of the pipe, withdrawing the pipe from the manhole, and directly transferring the collected sample to a labelled sample container. A PVC pipe was also used to collect samples from the Ala Wai Canal, which allowed a vertical composite sample from the entire depth of the canal (~0.5 to 1.7 m) to be collected. Three rounds of sampling were conducted from this monitoring network between November 2020 and January 2021 under mid- to high-tide conditions to evaluate the spatial distribution of sewage exfiltration in the Waikīkī and inland McCully-Moiliili areas. Additional sampling was conducted between February 2021 and December 2021 from three Waikīkī storm drains and two Ala Wai Canal locations to monitor the temporal variation in concentration levels of these compounds. Samples were collected during the global COVID pandemic, when the number of tourists staying in Waikīkī greatly fluctuated due to travel restrictions imposed by the State of Hawaiʻi. Fig. 1 shows the location of the five repetitively monitored sites along with two canal and groundwater monitoring locations in Waikīkī previously sampled in 2018 for pharmaceuticals [13].

The samples were analyzed for general water quality parameters (temperature, specific conductance, salinity, dissolved oxygen, pH and turbidity) in the field and frozen shortly after collection. Enzyme Linked Immunosorbent Assay (ELISA) test kits manufactured by Eurofins were used to measure the concentration of caffeine and carbamazepine, typically in batches of 80 samples at a cost of less than $8 per sample.

3.1 Pathway of sewage contamination entering coastal storm drain systems

Water levels were measured during this study using transducers under dry weather conditions in the Ala Wai Canal and in the coastal storm drain network underlying Waikīkī and the coastal portion of McCully-Moiliili. Water level fluctuations were identical in amplitude and timing as the tidal changes measured at the National Oceanic and Atmospheric Administration (NOAA) gauge in Honolulu Harbor (NOAA Station ID: 1612340).

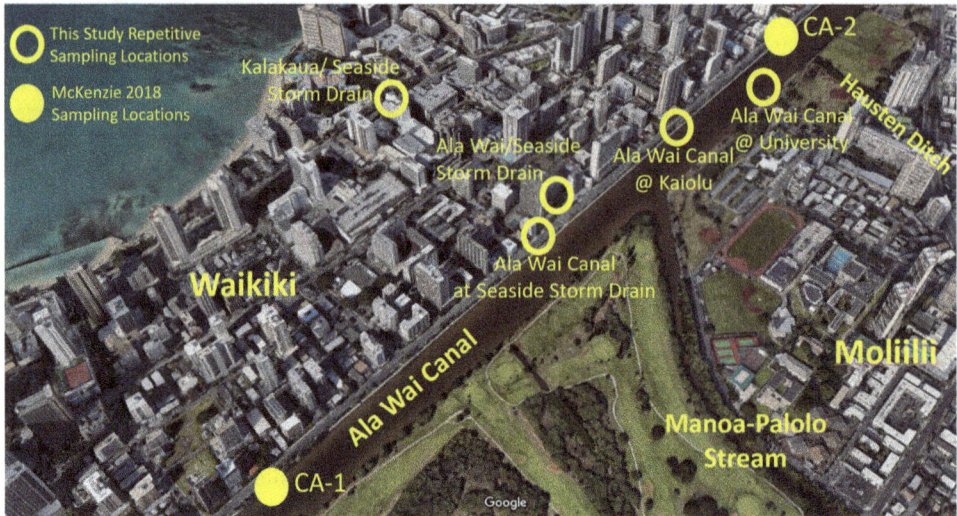

Figure 1:   Repetitively sampled locations in Waikīkī during this study (open circles) and two 2018 monitoring locations (closed circles) [13].

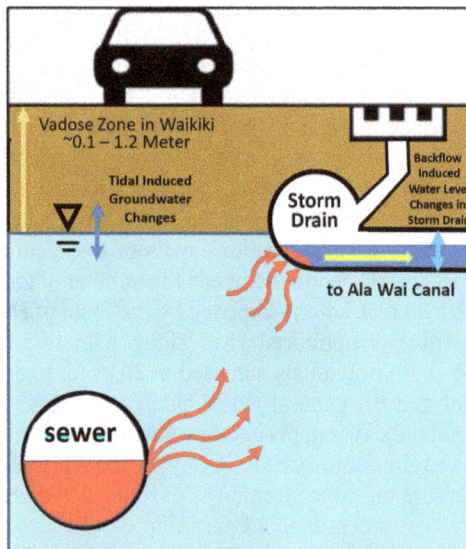

Figure 2:   Depiction of mechanism of how sewage exfiltration enters storm drain system in tidally influenced coastal areas on Oʻahu.

Previous studies have shown that leaking sanitary sewers can directly contaminate nearby leaking storm drains with untreated sewage during dry weather conditions [14]. Discharges can also enter the storm drain system indirectly, when groundwater contaminated by leakage from a nearby sewer enters the stormwater system. Fig. 2 depicts how leaks in sewer lines in coastal areas contaminate the surrounding brackish aquifer which can then enter cracks in the storm drain conveyance system during the lower portions of the tidal cycle when the groundwater levels are above the tidally driven water levels in the storm drain pipes. In Waikīkī, wastewater leaking outward from cracked pipes (exfiltration) in the study area migrates into the stormwater system, which acts as a very effective conduit to deliver leaking wastewater into the Ala Wai Canal. Based on the pattern of human consumption of caffeine (breakfast and dinner) and carbamazepine (taken orally twice per day), the concentration levels of these compounds in sewage tend to peak in mid-morning and early evening [10]. However, the caffeine and carbamazepine concentrations that leak into nearby storm drains have likely reached a quasi-steady state concentration in the contaminated groundwater during advective transport between the leaky section of sewer line and the cracks in the storm drain lines through which the contaminated groundwater enters.

Fig. 3 compares the salinities measured in the storm drain located at the intersection of Seaside and Kalakaua Boulevard in central Waikīkī with the salinities present in a storm drain located at the intersection of Seaside and Ala Wai Boulevard, directly adjacent to the Ala Wai Canal. The salinity in both the Waikīkī storm drain and the Ala Wai Canal increased by about 50% during high tide, as salt water from the Pacific Ocean is pushed landward into the storm drain with the tide. As the tide drops, the impact of brackish groundwater entering the storm drain system is seen in the falling salinities observed at both monitoring locations.

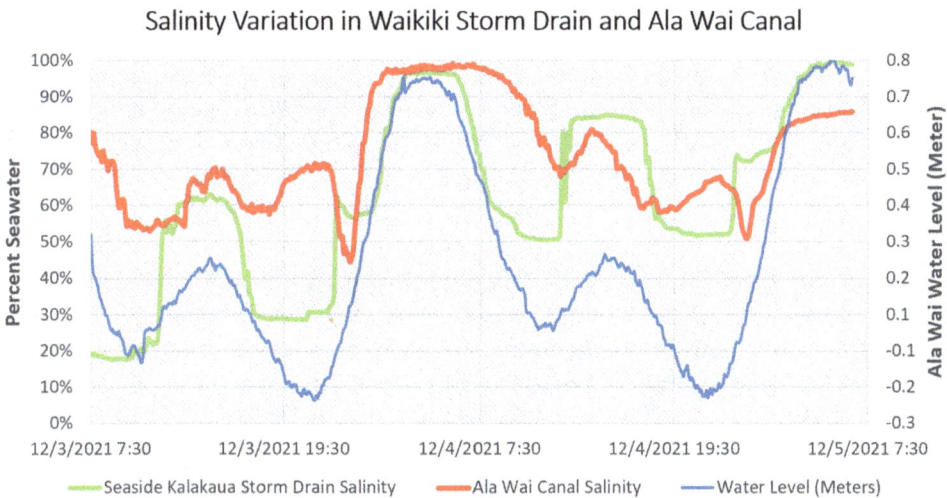

Figure 3:   Variation in Salinity and water levels measured in storm drains located in the middle of Waikīkī and adjacent to the Ala Wai Canal.

Fig. 4 shows the change in water levels measured in two connected storm drains located 410 m apart in central Waikīkī and adjacent to the Ala Wai Canal during a large winter rainfall event. The annual rainfall in Waikīkī is around 635 mm per year. A total of 180 mm of rainfall fell in Waikīkī between 30 December 2021 and 3 January 2022 (weather underground station KHIHONOL275 located in central Waikīkī along the Ala Wai Canal).

As can be seen in this graph, the water levels in the storm drains in central Waikīkī quickly rise as much as 0.9 m above the water levels in Ala Wai Canal during intense rainfall events, producing a gradient hydraulic gradient of as much as 0.0022 m/m that drives runoff from the center of Waikīkī towards the canal.

Waikiki Storm Drain Water Levels During Rainfall Event

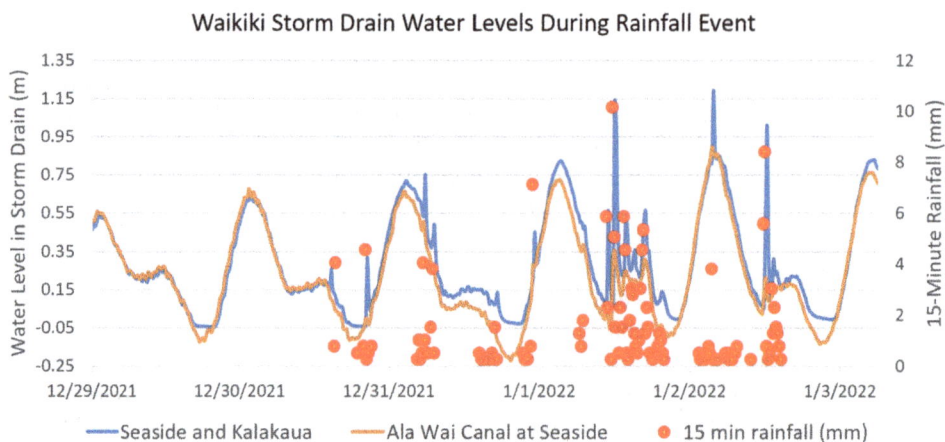

Figure 4: Water level variations in Waikīkī storm drains during a large rainfall event.

The average surface elevation of the majority of land in Waikiki is less than 1 m. Sea-level rise (SLR) is currently predicted to rise between 0.18 to 0.24 m from 1994 to 2014 sea levels by 2050 [15]. Groundwater inundation flooding of the coastal areas in this study area will occur contemporaneous with SLR related flooding as groundwater levels are lifted [16]. Fig. 4 illustrates how vulnerable Waikīkī is to future flooding events should a high-intensity rainfall event hit the area during a high or king tide. At high-tide, much of the capacity of the storm drain system in these coastal areas is filled with tidal water rendering the system unable to convey storm runoff towards the Ala Wai Canal or the Pacific Ocean.

4 SAMPLING RESULTS

The monitoring network of 70 manholes established in the Waikīkī and McCully-Moiliili areas were sampled a minimum of three times between November 2020 and January 2021 under upper mid- to high-tide conditions (typically greater than 0.55 m water levels) when tidal backflow partially fills the underlying storm drain conveyance system. Immunoassay methods were used to quantify the concentration levels of caffeine and carbamazepine present in the storm drain system and in Ala Wai Canal. Fig. 5 shows the storm drain and canal sites sampled (red dots), the layout and age of the sewer system in the area, the location of on-site sewage disposal systems in the area and the maximum carbamazepine concentration measured in the storm drain and canal samples (minimum of three samples) collected from each sampling site between November 2020 and January 2021.

Ponded water was present along the street curb under dry weather conditions near the storm drains sampled on Kaio'o Drive and Namahana Street during the 3 month sampling period (Fig. 5). The dry weather ponded water at these two sites had disappeared by April 2021 suggesting that the water present on the street was related to either a sewage line or water line leak at these locations. The elevated carbamazepine concentrations measured in samples collected from the Hausten ditch were likely related to work undertaken for the Moiliili Area Sewer Reconstruction project [17]. During the sampling work, contractors hired by the City and County of Honolulu (CCH) were replacing a section of deteriorated cast iron sewer line from 808 to 828 Hausten Street that was originally installed in 1935 and located adjacent to the inland end of the Hausten Ditch. According to a press release, CCH intended

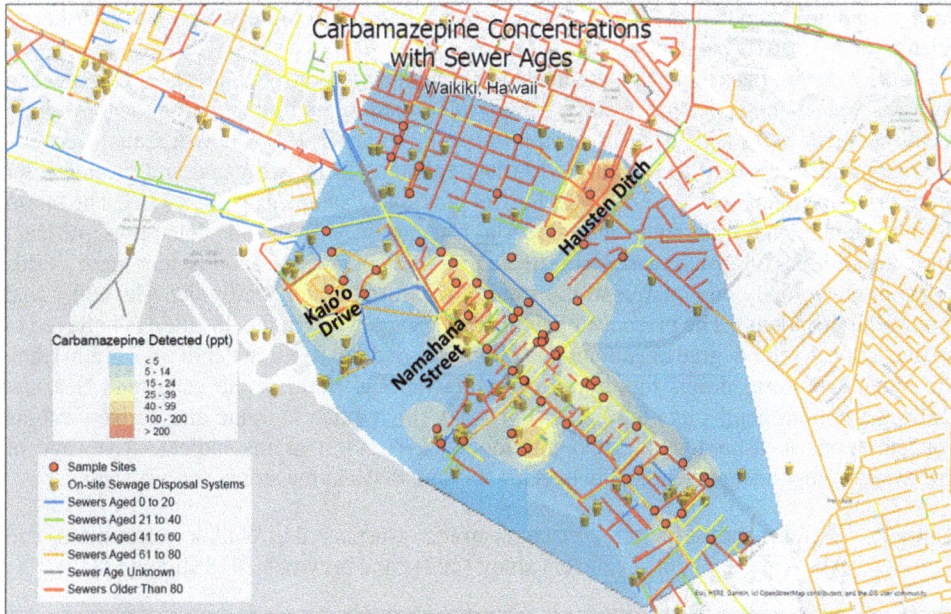

Figure 5:   Spatial distribution of the maximum carbamazepine concentration measured in Waikīkī and Moiliili-McCully storm drains between November 2020 and January 2021.

to bypass the sewer line continuously, for 24 hours, 7 days a week during construction using pumps and generators. Based on the elevated carbamazepine concentrations measured in the Hausten Ditch between November 2020 and January 2021, the sewer line being repaired was not successfully bypassed. A follow-on sample collected from the Hausten Ditch in early February 2021 after the sewer repair work was completed did not contain detectable levels of carbamazepine.

4.1 Time variation in caffeine and carbamazepine concentrations

The COVID pandemic had a dramatic impact on the Hawaiian tourist industry. Annual visitors to Oʻahu plummeted from over 6 million annual visitors between 2017 and 2019 to 1.5 million visitors in 2020 and 3.3 million visitors in 2021. The decline in the number of tourists was reflected in the volume of wastewater generated. The Sand Island WWTP processes roughly 60% of the wastewater generated on Oʻahu and covers the metropolitan Honolulu area, including the Waikīkī and McCully-Moiliili areas. The average daily sewage flow processed at the Sand Island WWTP declined about 12% from a 2017 average daily volume of 260 mld to an average daily volume of 228 mld in 2021.

Five monitoring locations (Fig. 1) (three in Waikīkī storm drains and two in the Ala Wai Canal) were repeatedly sampled at both low and high tide from February 2021 to December 2021. This time series sampling was conducted to evaluate whether the caffeine and carbamazepine concentrations present in Waikīkī storm drains and in the adjacent Ala Wai Canal varied: (1) as a function of the tidal cycle (i.e., high versus mid to low tide) and; (2) over the sampling period, when the 7 day average daily number of visitors to Waikīkī varied from around 6,000 to 30,000 visitors per day as a result of varying travel restrictions related to the COVID pandemic.

Table 2 summarizes the mean, median and maximum concentration levels of caffeine and carbamazepine along with the average salinity measured in the three monitored storm drain locations in Waikīkī and in the two monitoring locations in the Ala Wai Canal between February and December 2021 along with data collected from two canal and groundwater monitoring locations (CA-1 and CA-2, Fig. 1) previously sampled in 2018 [16]. The storm drain located inland of the intersection of Ala Wai and Seaside was impacted by significant intrusion of brackish groundwater as reflected by the low average salinity measured in this storm drain (roughly one quarter of the salinity measured at the nearby the Ala Wai Canal monitoring locations). The presence of significant groundwater at this location is believed to be due to the greater degree of submersion of this storm drain (bottom depth of -1.15 m) than the other two storm drains monitored (Seaside/Ala Wai drain at the canal and Seaside/ Kalakaua bottom depths of -0.24 and -0.45 m). The median caffeine and carbamazepine concentrations measured in the two interior Waikīkī storm drains indicate a sewage or greywater component presence of between 1.3% and 6.4% in the storm drain system.

Table 2:   Caffeine and carbamazepine concentrations measured in Waikīkī storm drains and Ala Wai Canal during current study (2021) and previous 2018 study [13].

Waikīkī Repetitive Sampling Locations	Caffeine (ppt)					
	Mean	Median	Max. Detect	Count	% Detect	% Seawater
Storm Drain: Kalakaua/Seaside	4,347	3,269	11,562	24	100%	58%
Groundwater Impacted Storm Drain: Ala Wai / Seaside	1,690	1,273	7,580	27	100%	21%
Storm Drain: Ala Wai Canal at Seaside	677	398	2,793	27	81%	74%
Ala Wai Canal at Kaiolu	543	207	1,735	25	88%	80%
Ala Wai Canal at University	287	82	1,439	24	71%	79%
CA-1 and CA-2 Surface Water[1]	1,091	1,010	2,700	8	75%	NR
CA-1 and CA-2 Groundwater[1]	986	1,045	1,600	8	88%	NR
Source	Carbamazepine (ppt)					
	Mean	Median	Max. Detect	Count	% Detect	% Seawater
Storm Drain: Kalakaua/Seaside	28	12	136	29	93%	58%
Groundwater Impacted Storm Drain: Ala Wai / Seaside	35	32	102	33	100%	21%
Storm Drain: Ala Wai Canal at Seaside	7	2	27	31	67%	74%
Ala Wai Canal at Kaiolu	6	0	44	30	37%	80%
Ala Wai Canal at University	6	0	36	29	48%	79%
CA-1 and CA-2 Surface Water[1]	44	0	150	8	75%	NR
CA-1 and CA-2 Groundwater[1]	83	105	130	8	75%	NR
NR = Not Reported						
[1]Surface and Groundwater Samples Collected in 2018 [13]						

The highest mean/median caffeine concentrations were measured in the storm drain in central Waikīkī while slightly higher carbamazepine concentrations were measured in the groundwater impacted storm drain at the intersection of Ala Wai and Seaside. The caffeine concentrations measured in the storm drain along the Waikīkī side of the Ala Wai Canal and at the canal monitoring location at Kaiolu Street were higher than the caffeine concentrations measured at the inland monitoring location across the canal at the end of University Avenue.

The median caffeine concentrations measured in the groundwater impacted storm drain in 2021 were similar to caffeine levels measured in groundwater in 2018 (1,273 versus 1,045 part per trillion (ppt)) while the caffeine concentrations measured in the Ala Wai Canal in 2021 were significantly lower than 2018 levels (82-398 versus 1,100 ppt). The median carbamazepine concentrations measured in the groundwater impacted storm drain in 2021 were less than a third of the carbamazepine levels measured in groundwater in 2018 (32 versus 105 ppt). This difference in concentration levels measured in 2018 and 2021 is consistent with the higher throughput of sewage in the Waikīkī area (and corresponding exfiltration) in 2018 from the 6 million annual visitors to Oʻahu that year compared to the 1.5 to 3.3 million visitors in 2020 and 2021.

The median concentration of caffeine measured in the storm drains at Seaside/Kalakaua and the groundwater impacted storm drain at Seaside/Ala Wai (Fig. 1) were higher at low tide than high tide (4,160 versus 2,750 ppt and 1,850 versus 790 ppt, respectively). The median concentration of carbamazepine measured in the groundwater impacted storm drain at Seaside/Ala Wai was higher at low tide than high tide (41 ppt versus 20 ppt) while the carbamazepine concentrations measured at the Seaside/Kalakaua storm drain at low and high tide was similar (12 ppt versus 15 ppt). It is important to note that the trace levels of carbamazepine detected during this study are significantly lower than ecotoxicological levels such as the predicted no-effect concentration of 25,000 ppt and the critical environmental concentration of 349,496 ppt determined for this pharmaceutical compound [3].

Fig. 6 shows the variation in caffeine concentrations measured in the storm drain at Seaside/Kalakaua (yellow dots) and in the groundwater impacted storm drain at Seaside and Ala Wai (red dots) along with the average 7 day daily arrival of visitors to Oʻahu in 2021.

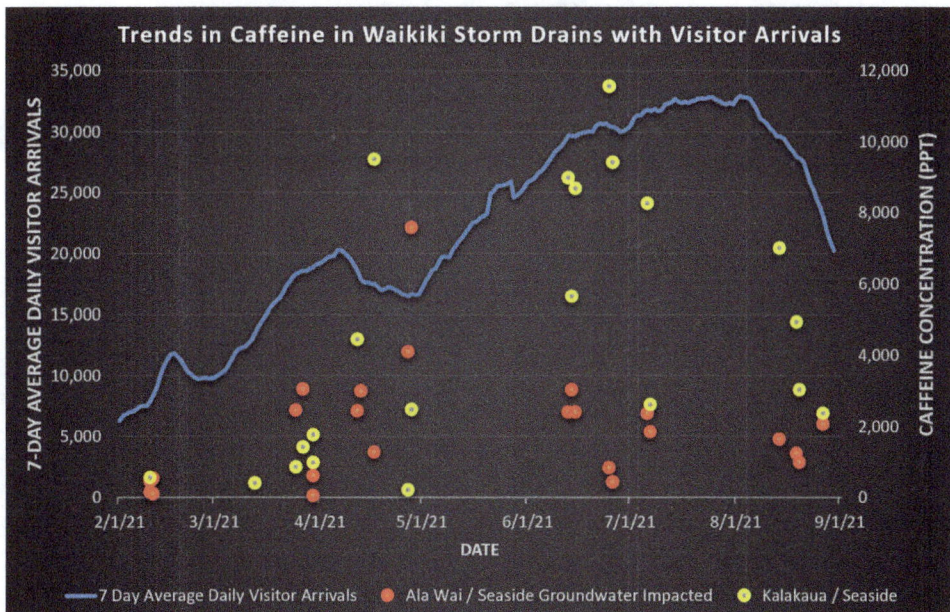

Figure 6:   Trend in caffeine concentrations in Waikīkī storm drains and visitor arrivals to Oʻahu.

The caffeine concentrations measured at the Seaside/Kalakaua monitoring locations increased in the middle of the year as tourist arrivals rose then started to decline at the end of the summer as tourist arrivals declined. The trend in caffeine concentrations at the groundwater impacted storm drain at Seaside/Ala Wai are broadly similar but more variable and at overall lower concentrations suggesting the existence of additional input of caffeine to the storm drain at Seaside/Kalakaua than is present in the sewage contaminated groundwater underlying Waikīkī.

5 CONCLUSIONS

This study has shown that sampling of storm drains in low-elevation coastal portions of urban areas such as the city of Honolulu offers an attractive, inexpensive alternative for locating general areas of sewage exfiltration to the traditional exfiltration location methods used. The immunoassay methods used to quantify the concentration of caffeine and carbamazepine present in water in the storm drains and in Ala Wai Canal cost less than $8 per analysis when samples are run in baches of 80 samples. In this study, the pharmaceutical compounds caffeine and carbamazepine were used to identify areas of sewage exfiltration in the Waikīkī and McCully-Moiliili areas in urban Honolulu. Alternative pharmaceutical compounds or artificial sweeteners such as sulfamethoxazole or sucralose would also likely be effective at delineating areas of sewage exfiltration in urban coastal settings.

REFERENCES

[1]   American Society of Civil Engineers (ASCE), 2019 Hawai'i Infrastructure Report Card, 2019. www.infrastructurereportcard.org/hawaii.
[2]   Amick, R.S. & Burgess, E.H., Exfiltration in sewer systems. EPA/600/R-01/034, December 2000.
[3]   Wilkinson, J.L. et al., Pharmaceutical pollution of the world's rivers. *PNAS*, **119**(8), e2113947119, 2022. DOI: 10.1073/pnas.2113947119.
[4]   Mitchell, D.C., Knight, C.A., Hockenberry, J., Teplansky, R. & Hartman, T.J., Beverage caffeine intakes in the U.S. *Food Chem. Toxicol.*, **63**, pp. 136–142, 2014.
[5]   Rybak, M.E., Sternberg, R., Pao, C.-I., Ahluwalia, N. & Pfeiffer, C.M., Urine excretion of caffeine and select caffeine metabolites is common in the US population and associated with caffeine intake. *J. Nutr.*, **145**(4), pp. 766–774, 2015.
[6]   Meppelink, S.M., Kolpin, D.W., Lane, R.F., Iwanowicz, L., Zhi, H. & LeFevre, G., Water-quality data for a pharmaceutical study at Muddy Creek in North Liberty and Coralville, Iowa, 2017–2018. DOI: 10.5066/P9WOD2XB.
[7]   Luo, Y., Guo, W., Ngo, H.H., Nghiem, L.D., Hai, F.I., Zhang, J., Liang, S. & Wang, X.C., A review on the occurrence of micropollutants in the aquatic environment and their fate and removal during wastewater treatment. *Sci. Total Environ.*, **473**, pp. 619–641, 2014.
[8]   Benotti, M.J. & Brownawell, B.J., Distributions of pharmaceuticals in an urban estuary during both dry- and wet-weather conditions. *Environmental Science and Technology*, **41**(16), pp. 5795–5802, 2007.
[9]   Hai, F., Yang, S., Asif, M., Sencadas, V., Shawkat, S., Sanderson-Smith, M., Gorman, J., Xu, Z.-Q. & Yamamoto, K., Carbamazepine as a possible anthropogenic marker in water. *Water*, **10**(2), p. 107, 2018. DOI: 10.3390/w10020107.
[10]   Shelton, J.M., Kim, L., Fang, J., Ray, C. & Yan, T., Assessing the severity of rainfall-derived infiltration and inflow and sewer deterioration based on the flux stability of sewage markers. *Environ. Sci. Technol.*, **45**, pp. 8683–8690, 2011.

[11] Hawai'i Department of Health (HDOH), 2018 Groundwater status report. Appendix H: Assessing the presence and potential impacts of pharmaceutical and personal care products (PPCPs) on groundwater and drinking water, 2019.

[12] Hawai'i Department of Business, Economic Development (HDBED), Tourism data warehouse, visitor statistics, 2022.

[13] McKenzie, T., Habel, S. & Dulai, H., Sea-level rise drives wastewater leakage to coastal waters and storm drains. *Limnology and Oceanography Letters*, **6**, pp. 154–163, 2021. DOI: 10.1002/lol2.10186.

[14] Sercu, B., Van De Werfhorst, L.C., Murray, J.L.S. & Holden, P.A., Sewage exfiltration as a source of storm drain contamination during dry weather in urban watersheds. *Environ. Sci. Technol.*, **45**(17), pp. 7151–7157, 2021.

[15] IPCC, 2021: *Climate Change 2021: The Physical Science Basis. Contribution of Working Group I to the Sixth Assessment Report of the Intergovernmental Panel on Climate Change*, Cambridge University Press, in press.

[16] Habel, S., Fletcher, C.H., Rotzoll, K. & El-Kadi, A.I., Development of a model to simulate groundwater inundation induced by sea-level rise and high tides in Honolulu. *Hawaii Water Res.*, **114**, pp. 122–134, 2017.

[17] Akinaka & Associates, Moiliili-Kapahulu sewer rehabilitation/reconstruction. Final environmental assessment, 2011.

DATA INTEGRATION, HARMONIZATION AND PROVISION TOOLKIT FOR WATER RESOURCE MANAGEMENT AND PREDICTION SUPPORT

GEORGIOS VOSINAKIS*, EVANGELOS MALTEZOS, MARIA KROMMYDA,
ELEFTHERIOS OUZOUNOGLOU & ANGELOS AMDITIS
Institute of Communication and Computer Systems (ICCS), Greece

ABSTRACT

Timely and reliable information is critical to organizations managing water resources. Drinking water is one main source of risk when its safety and security is not ensured. Early prediction and mitigation of such risks relies on prediction models that depend on live and historical data. Such data are quite heterogenous in nature, including sensor measurements, satellite imagery and radar readings, unmanned aerial vehicle (UAV) images and videos as well as results of prediction algorithms (flood risk, oil spills etc). AQUA3S is an EU funded project which combines novel technologies in water safety and security, aiming to standardize existing sensor technologies complemented by state-of-the-art detection mechanisms. Sensor networks are deployed in water supply networks and sources, supported by complex sensors for enhanced detection. Sensor measurements are supported by videos from UAVs, satellite images and social media observations from the citizens that report low-quality water in their area also creating social awareness and an interactive knowledge transfer. Semantic representation and data fusion provides intelligent decision support system (DSS) alerts and messages to the public through first responders' mediums. This study presents the data ingestion, integration and harmonization platform that was developed to support the systems of the project, consisting of the necessary APIs, to ingest data, a harmonization layer and a data store layer The data is harmonized and indexed using the NGSI-LD model to make sure information can be indexed and served both is real time through a live context broker, as well as in the form of historical time series through a dedicated historical data service. The data store layer includes provisions for the storage of annotated binary files (images, videos, etc.) as well as georeferenced map layers following OGC protocols such as web feature service (WFS), web map service (WMS), and web coverage service (WCS).

Keywords: water safety, water management, WFS, WMS, WCS, decision support, NGSI, linked data, water resource management, digital twins.

1 INTRODUCTION

The management of water networks and water reservoirs is a critical aspect for any society. Water management combines a variety of fields from resource management to predicting and defusing crisis situations. Timely information and readily available, well indexed data are critical to informed decisions during a quickly evolving crisis, such as water contamination incidents or flooding of inhabited areas due to adverse weather conditions. At the same time recording and monitoring of land use in areas around critical water sources is vital to the provision of adequate clean water. An increasing number of algorithms and prediction systems have been suggested that use a variety of diverse data sources. In this context several approaches have been suggested utilising sensor measurements [1], [2] as well as satellite imagery [3], [4] and even citizen information available through social media [5], [6].

AQUA3S [7] is an EU funded project that aims to combine novel technologies in water safety and security and standardize existing sensor technologies complemented by state-of-the-art detection mechanisms, providing water suppliers with timely information and

---

* *ORCID: http://orcid.org/0000-0003-2029-8409*

WIT Transactions on The Built Environment, Vol 208, © 2022 WIT Press
www.witpress.com, ISSN 1743-3509 (on-line)
doi:10.2495/FRIAR220071

assessments of evolving situations. In this context the harmonization and availability of data in a ready to be consumed format is integral to the success of the project and the operation of the platform. Based on an analysis of the available data types, as well as user and technical requirements gathered during the initial phase of the project, a data ingestion and harmonization layer was designed and developed, together with a data store layer and the necessary API services that allow for the indexed storage and request of data by analytical modules, data visualization modules, as well as authorized data consumers directly.

The available data belongs to two main categories, primary data, available through a variety of sources and legacy sensor systems and secondary data, produced by the analytical modules of the project platform. After ingestion, data are stored and indexed to facilitate availability. This is achieved through a harmonization layer. Data were harmonized using Fiware compatible smart data models [8] and following the FIWARE NGSI-LD standard [9]. Several new data models, used in indexing the data types of the project were developed by the consortium partners (e.g. for satellite imagery, social media data, risk management). The main focus was given to providing a solution that would be easy to extend and adapt to new data types, as well as configure to new requirements.

Fig. 1 shows an example of an NGSI device entity in JSON-LD format. The entity includes a unique URN ID for the entity (line 3). The linked data [10] format of these data models allows (through their URN ID to establish connections between entities. In this case a "relationship" attribute (lines 40–44) connects this sensor to a specific real-world asset (water reservoir).

```json
1 ▾ {
2      "@context": "https://schema.lab.fiware.org/ld/context",
3      "id": "urn:ngsi-ld:Device:Sensor_001",
4      "type": "Device",
5 ▾    "controlledProperty": {
6          "type": "Property",
7          "value": "freeChlorine"
8      },
9 ▾    "value": {
10         "type": "Property",
11         "value": 0.1405,
12         "observedAt": "2022-02-28T16:10:00.000Z",
13         "unitCode": "M1"
14     },
15 ▾   "name": {
16         "type": "Property",
17         "value": "0121 Chlorine Residual"
18     },
19 ▾   "dateLastValueReported": {
20         "type": "Property",
21 ▾       "value": {
22             "@type": "DateTime",
23             "@value": "2021-12-16T14:05:00.000Z"
24         }
25     },
26 ▾   "deviceState": {
27         "type": "Property",
28         "value": "Green"
29     },
30 ▾   "location": {
31         "type": "GeoProperty",
32 ▾       "value": {
33 ▾           "coordinates": [
34                 34.938792,
35                 18.326333
36             ],
37             "type": "Point"
38         }
39     },
40 ▾   "controlledAsset": {
41         "type": "Relationship",
42 ▾       "object": {
43             "urn:ngsi-ld::reservoir-NorthWest-100"
44         }
45     }
46 }
```

Figure 1:   Example of an NGSI-LD device entity, used to map sensors (in this case a residual chlorine sensor).

A central context broker allows for the storage and retrieval of data in NGSI-LD compatible format. With the harmonization of available data to the NGSI-LD standard and the indexing of binary data (e.g. satellite images) using the data models developed, the integration platform can be connected to multiple services and expanded in the future to include more data sources, or new data types and can support the data requirements of any client module (analytical, visualization, etc.) that might be developed in the future based on the NGSI-LD standard.

The platform has been designed to accommodate the needs of the AQUA3S project but, by offering basic data harmonization and storage services it can be considered "data type agnostic" and can be easily adapted to other fields, beyond water quality and management..

2 DATA TYPES CONSIDERED

The data ingestion architecture developed has been designed based on the data types available to the project. The aim of this layer is to collect data from multiple heterogenous sources that will be harmonized, processed and indexed. Primary data, available from various sources comprise:

- Live sensor data;
- Sensor measurements from legacy SCADA systems;
- Satellite imagery, with accompanying metadata;
- Social media data, with accompanying metadata;
- Call complaint data from water suppliers' call centres;
- Water Network data, available as EPANET files;
- CCTV and Drone images and footage, with accompanying metadata.

While secondary data, produced by analytical modules within the AQUA3S platform comprise:

- Results of social media data analysis;
- Results of satellite image analysis;
- Geo-referenced map layers with accompanying metadata;
- Crisis classification analysis results.

The aim of our ingestion platform is to collect data from multiple sensors and other sources and has a central role in the integration of heterogeneous sources of data. It essentially consists of a logical bus implementing a publish–subscribe mechanism aiming to advertise the field devices data and metadata as resources as well as the relevant adapters for translation of the proprietary sensor data to a harmonized data format implemented by a middleware layer.

3 SYSTEM ARCHITECTURE

The architecture of our system consists of two distinct layers:

- A data ingestion and harmonization layer, responsible for ingesting data into the system, as well as indexing and aligning data received with the required data models.
- A data store layer, responsible for storing indexed data and the accompanying APIs that allow for the effective retrieval of data in a structured manner.

A general overview of the system architecture is shown in Fig. 2.

WIT Transactions on The Built Environment, Vol 208, © 2022 WIT Press
www.witpress.com, ISSN 1743-3509 (on-line)

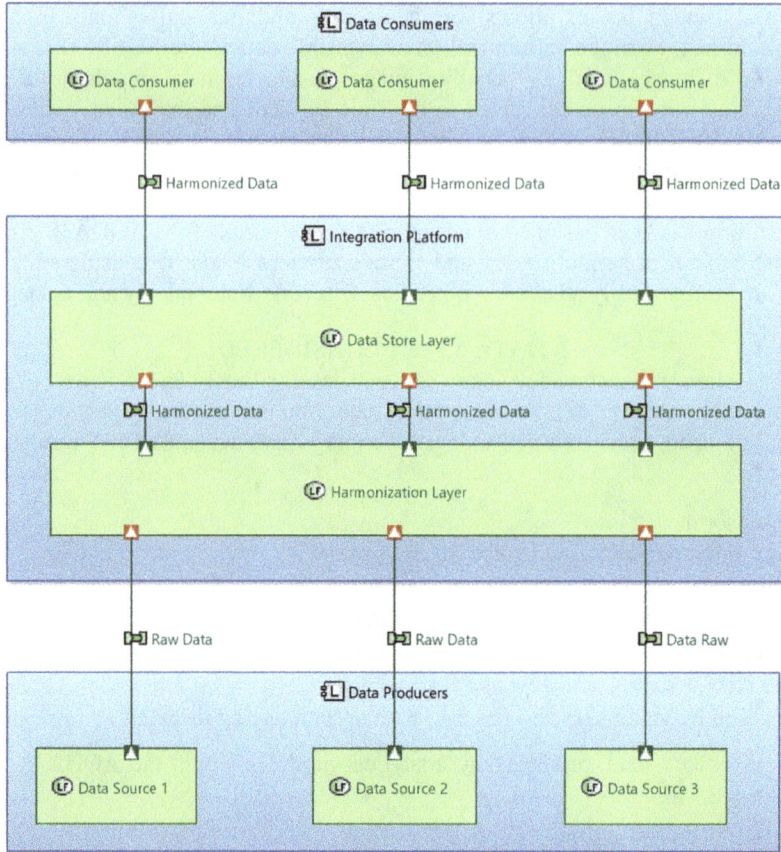

Figure 2: The general block diagram of the system architecture, showing the Integration platform with its distinct layers.

3.1 Data store layer

The data store layer consists of a context broker, holding information in JSON-LD format on sensor measurement and analysis results. Binary files are stored in a series of binary storage modules while a corresponding entity is created within the context broker, recording important information on the binary file, as well as its location in the binary storage. In this context, the broker is the central module of the structure holding the current, up-to-date image of the entire system. It provides, directly or indirectly through links, all the information ingested by data consumers and analytical modules.

The core of the data store layer is based on four main, open-source tools:

- An Orion-LD context broker [11] – It stores and serves the current state of the system (latest values) using the JSONLD format and adhering to Fiware smart data models.
- Cygnus-LD historical enabler [12] – Paired with a historical service API, developed within the project to serve queries to the historical database, it stores and serves historical information and time series of historical values.
- A GeoServer [13] – It stores and serves georeferenced images and data.

- A WebDAV server [14] – It stores and serves large binary files (e.g. images, videos and documents) and allows the distribution and collaboration on files.

Orion-LD allows for the creation of entities (e.g. corresponding to specific sensors) in JSON-LD format. Entities are indexed based on a unique entity ID and categorized based on entity type. Entities can be queried through HTTP requests based on their type or unique entity ID. This allows data consumers to receive the current status of a specific entity or the current status of all entities of a specific type (e.g. all sensors). Entities can be updated with new values. Orion-LD will hold and provide, when queried, the current, latest status of each entity monitored. When an entity is updated with a new value (e.g. a new measurement) Orion-LD will replace the previous with the latest. Orion-LD does not store previous values of entities, this is achieved through the Historical Service.

The Historical Service is responsible for storing and providing historical information of all entities held in Orion-LD. It consists of two components, Cygnus-LD, which is responsible for storing the historical information and the Historical Service API which provides the historical data to the data consumer modules through HTTPS requests. For every entity within Orion-LD a subscription is being created to the specific entity, establishing Cygnus-LD as the endpoint. Whenever the entity is updated in Orion-LD, the broker sends an automatic notification, including the previous status of the entity to Cygnus-LD. Cygnus-LD stores the status to a Postgres Database, indexed by its timestamp. Whenever a consumer module needs a timeseries of historical values for any entity, a request can be sent to the Historical Service API that will return a list, containing the historical values. Requests can be made using the specific entity ID and can either specify a time interval (starting date/time, ending date/time) or get values from a specific time/date to the present. Cygnus-LD is an open-source solution, but while it records data to a historical database, it does not provide an API to retrieve this data. This is the reason the Historical API was developed within the project to provide a Fiware compatible API service for the querying and provision of historical time series. Fig. 3 shows a general diagram of the historical service's architecture.

3.2 Integration and harmonization layer

A series of adaptors, developed in the Python 3.6 programming language [15] have been developed to facilitate the ingestion and harmonization of available data to the data models used within the project.

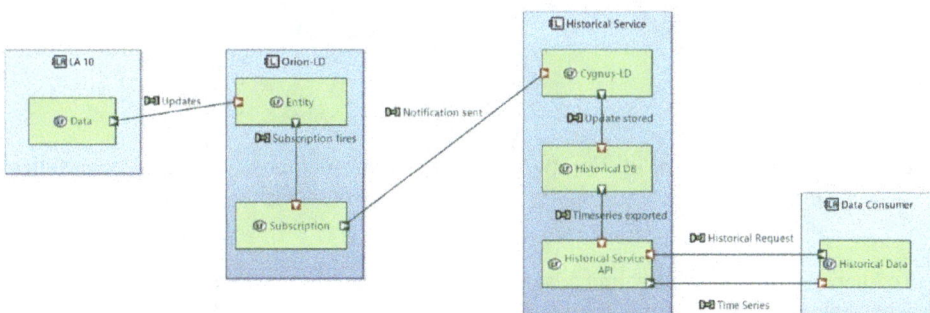

Figure 3:   A block diagram of the historical service, comprising the Cygnus-LD connector (open source) and the Historical Service API developed within the AQUA3S project.

The harmonization process is performed in two ways, depending on the specific requirements of each data source:

- Client scripts – Installed as embedded service within the data producing modules and transmitting data in a decentralized manner.
- Services that query available APIs or servers, and centrally collect information from available sources.

Clients are developed as lightweight scripts. They are meant to be installed in either AI units or single board computers used by independent sensors without overloading their operation. In the case of clients, the harmonization layer, which aligns data with the NGSI-LD standard, is included with the client so that data transmitted by the device is NGSI-LD compatible thus making the device an independent, integrated data transmitter. The client is installed on the device and the script is run either at regular intervals to receive updated measurements (in the case of sensors) or when new data is available for transmission (in the case of analytical services).

Centralized ingestion services incorporate various modules depending on the required communication protocols used to query data. All such services are available as Docker containers that run centrally on the AQUA3S platform. Depending on the availability and requirements for data, they run either periodically (in the case of sensors and legacy SCADA systems) or on demand when an analytical service needs new data to consume. Fig. 4 shows a diagram of the two harmonization methods.

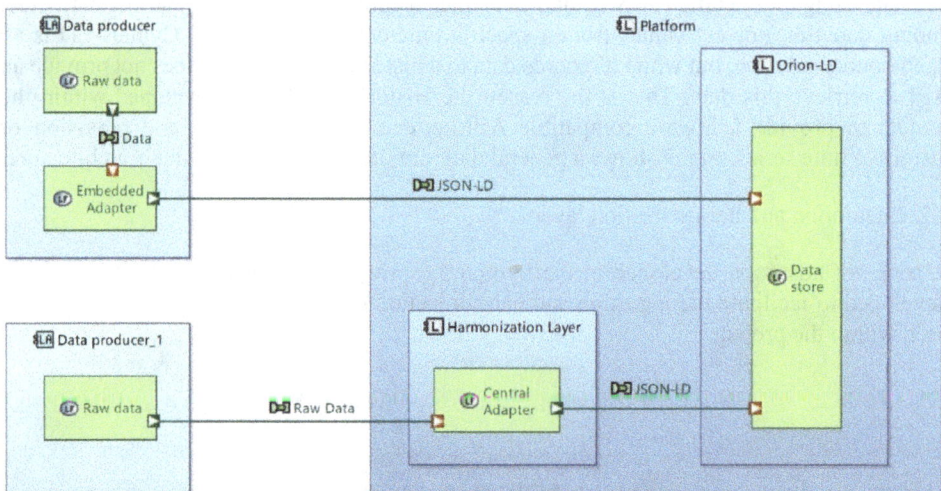

Figure 4:   Block diagram of the two harmonization methods, through decentralized embedded adapters (clients) and a central harmonization service.

To guarantee the reusability and adaptability of the adaptors used within the project to the various heterogenous data types ingested two Python 3 libraries were developed:

- Harmonization library: Implements a class of JSON parsers that include methods to ingest raw data and parse them into the correct JSON-LD structure, according to the corresponding data model.

- Connection library: Implements a class of connectors that handle connections to Orion-LD, providing methods to create, update or delete entities in Orion-LD, perform diagnostic checks and handle subscriptions and notifications.

4 DATA HARMONIZATION, STORING AND INDEXING

Data considered can be divided, based on their availability and format, into four basic categories: live data feeds, recorded data, georeferenced map layers and images and binary files. To keep our architecture "data type agnostic" in order to guarantee expandability to different data sources we developed a generalized strategy in handling each distinct data type.

4.1 Live data feeds with an embedded client

Live data feeds originate either from sources external to the project platform (like sensors that provide a live feed of readings) or internal sources (like analytical tools that process available data to produce analysis results).

Sensors provide periodical measurements at medium or high frequencies (1 Hz to 1/60 Hz). These measurements are either stored locally in a (usually unstructured) JSON file or made available through a socker or API. Sensors are installed at the end users' locations, where they need to connect to a wi-fi or mobile network to transmit data. Analytical tools can either operate automatically, to provide results in a periodical manner similar to sensors, or run on demand, to produce results on an ad hoc basis.

In both cases an on-board Python service was developed that will be installed on the sensor or analytical tool as an embedded client. In the case of a sensor, the service polls the measurements provided by the sensor at specific internals, harmonizes them to the JSON-LD data model and transmits them to the Orion-LD context broker. In the case of an analytical tool, the client is accessed by the tool itself when data is ready to be transmitted. Analytical tools that have been integrated using this method include:

- Satellite imagery analysis modules, processing satellite images and analyzing water resource related risks and crisis.
- Social media modules, including a crawler that collects social media posts relevant to water supply issues and an analytical tool that uses the posts as input.
- A crisis analysis module, combining several data sources and producing risk estimations on water supply and flooding.

Communication with the platform is through a wi-fi or mobile connections. Fig. 5 shows a diagram of this structure.

This decentralized solution was selected for independent sensors, based on common IoT practice, to make sure the ingestion service is expandable and adaptable. A configuration file provided with the adapter installed on the sensor allows end user to successfully configure the adapter to specific sensors.

When a new sensor needs to be added all the end user has to do is install the adapter on the sensor's hardware and update the configuration file with the new sensor's information. When the service connects to a network it will transmit measurements to the configured Orion-LD context broker.

The available configurations allow the users to select what happens in case of connection loss. The sensor can store measurements locally and transmit them when connection is established (to avoid data loss), or it can send only the latest measurement to ensure speed of service.

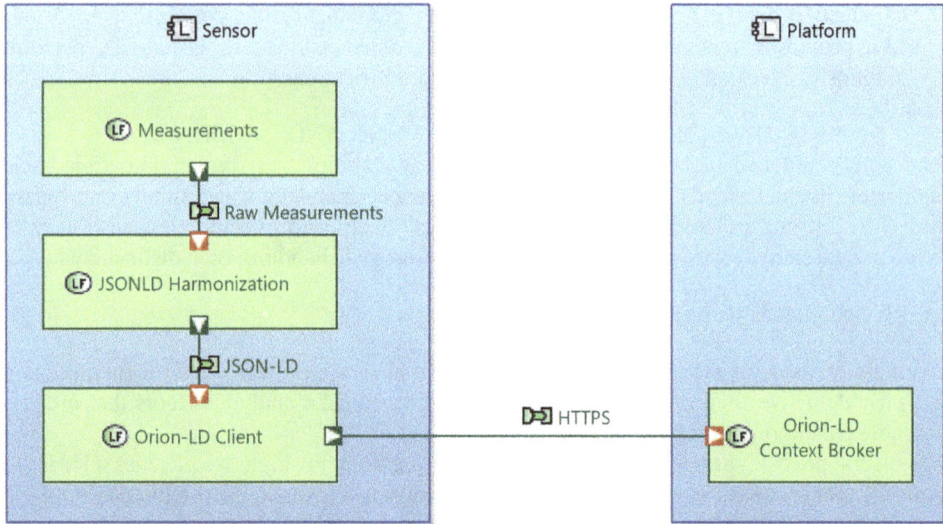

Figure 5: Block diagram of the embedded ingestion and harmonization client, using a sensor as an example.

4.2 Near real time and other recorded data

This category comprises a variety of data types that are available irregularly in the form of CSV or TXT files. These files, in the case of the AQUA3S project include bulk measurements provided by legacy SCADA systems, as well as call complaint records from water suppliers' call centres and EPANET [16] files that map water networks in a text format.

Data are collected from SCADA systems periodically and are made available to the project in batches, at specific time intervals. As a result, such data are near-real-time. Every time a batch is available a CSV file with the measurements is exported and uploaded to a dedicated sFTP server. CSV files generated in this manner have a filename starting with the unique sensor ID and include a column of measurements and a column with the corresponding timestamps. In the case of call complaints data are uploaded periodically by end users to a sFTP server in the form of csv files. The files list anonymized information on each call and its subject.

For sensor measurements, the ingestion service consists of a Python 3 application that acts as a server-side component that has the role of connecting to multiple sFTP sites, reading a list of asset files and reading a number of different measurement csv files. This functionality is based on periodic tasks running which connect and retrieve csv files. The service manages a list of queues on a per sensor/device basis. Queues contain the retrieved sensor measurements from the sFTP sites. The previous periodic tasks feed these queues.

It also handles a periodic task that reads one by one sensor measurements from the previous queues, performs harmonisation to the NGSI-LD data model and, via a rest client, uploads them to the Orion-LD context broker. To guarantee the expandability of the ingestion service each partner maintains within the corresponding sFTP server a configuration CSV file that lists the sensors providing measurements and their basic information (e.g. ID, name, location). Fig. 6 provides a general block diagram of this process.

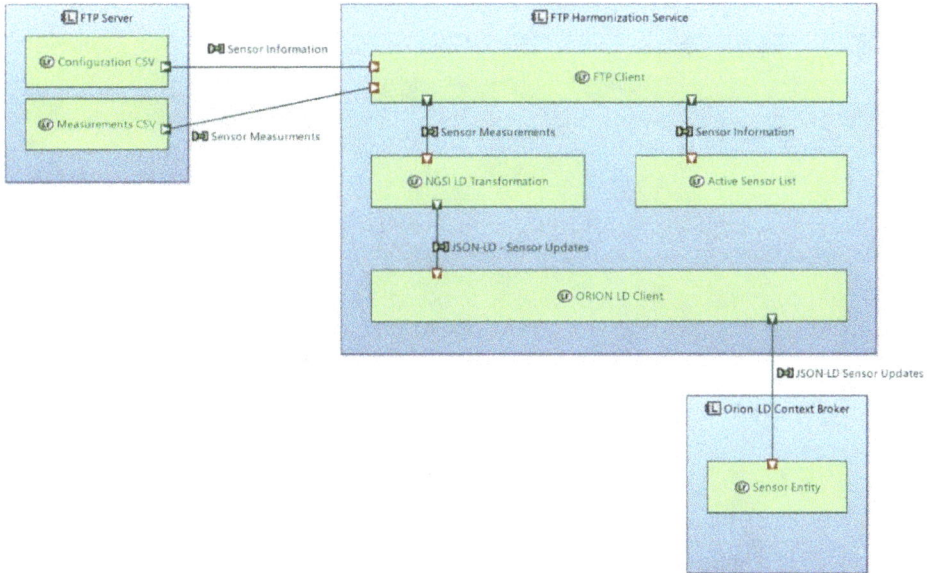

Figure 6: Ingestion and harmonization service for legacy SCADA systems.

When a new sensor needs to be added all the end user must do is update the configuration file with the new sensor's information. The service periodically ingests the configuration file and then automatically retrieves CSV files corresponding to those sensors. If a sensor is removed from the configuration file, the service will stop ingesting files from the corresponding sensor.

Call complaint records are integrated in a similar manner. An sFTP client periodically polls an sFTP server for new uploaded files. If new files are found, they are sent to the harmonization service which harmonizes data to the data model and uploads them to Orion-LD.

The analytical tool receives call complaints from the context broker and, after analysing them, sends the analysis results to its own harmonization service that publishes analysis data to the Orion-LD context broker.

4.3 Georeferenced map layers

This category includes satellite data comprising georeferenced satellite photographs that are acquired from online, open-source satellite data hubs (e.g. The Copernicus Hub), as well as the products of analyses produced by analytical modules developed within the project. The available data include georeferenced photographs and JSON files that include information of the performed analysis, as well as metadata on the photographs and their source.

After processing in the analytical service of the platform, photographs are uploaded to a GeoServer instance loaded on the platform. Information on the analysis performed, as well as metadata, including the photograph's location in the GeoServer are then sent to the harmonization service, which is responsible for harmonizing the data to the Orion-LD, Fiware compatible data model. The harmonization service includes a JSON-LD parser and an Orion-LD client and is attached to the analytical service. It is called ad hoc by the analytical service when new data is available for uploading (Fig. 7).

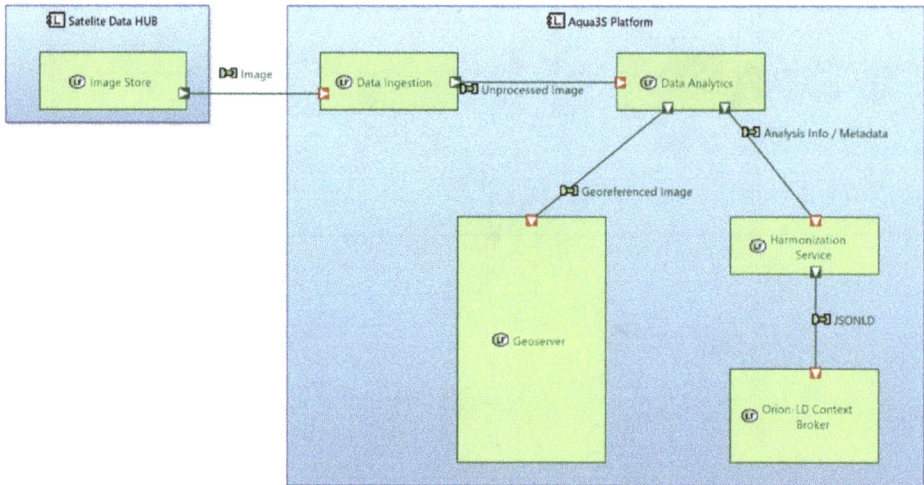

Figure 7: Ingestion and harmonization of satellite images.

4.4 Binary files

Large binary files include photographs and videos from drones or CCTV cameras operating at the water supplier's facilities or at important bodies of water. Such files are saved in the WebDAV server provided, while any connected metadata are harmonized and saved in the Orion-LD context broker, along with the location of the relevant file in the WebDAV server. If an end user, or analytical/visualization tools need to access the file they can query the broker with the relevant information and receive a link to the file in WebDAV. Fig. 8 shows a diagram of this architecture.

5 CONCLUSION

The ingestion platform is overall a lightweight and easy to install application based on open-source tools and unique integration solutions developed in Python 3. It is focused on being easily adaptable and expandable to different frameworks, providing libraries that can be reused in different applications as well as targeted configuration files that can be used to adjust the platform's operations in an intuitive manner.

The harmonization layer consists of a library used to parse Fiware compatible JSON-LD payloads, incorporating methods that can be used to expand the library to new data models if the need arises. It also incorporates an Orion-LD client that manages entity creation, updates, deletes and queries, as well as the handling of multitenancy and subscriptions.

Based on the heterogenous data that were available to the project in many different formats, the focus was given to providing an easily configurable and adaptable solution. With the harmonization of available data to the NGSI-LD standard and the indexing of binary data (e.g. satellite images) using the data models developed, the integration platform can be connected to multiple services and expanded in the future to include more data sources, or new data types and can support the data requirements of any client module (analytical, visualization etc) that might be developed in the future based on the NGSI-LD standard. The platform can be considered "data type agnostic" and can be easily adapted to other fields, beyond water quality and management.

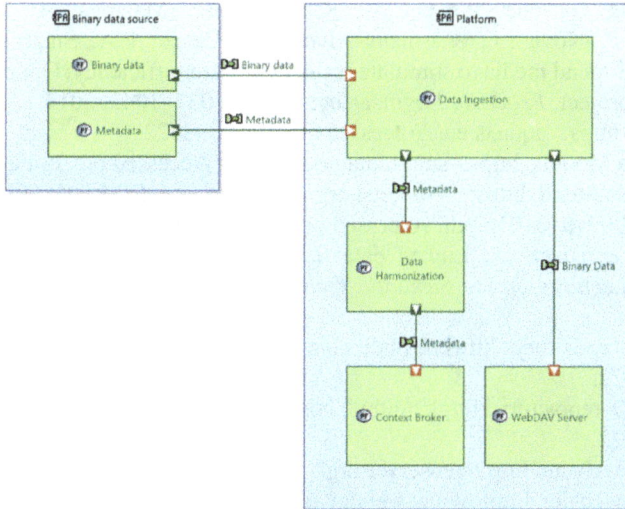

Figure 8: Schematic of binary data ingestion process.

Several steps have already been taken and will continue to be taken to allow for the fast, effective, and easy deployment of the platform and facilitate its connection with new data providers and consumers. Such steps include the standardization of development libraries that support data creation and queries as well as generalizations on the queries necessary for data retrieval by data consumers or analytical tools.

ACKNOWLEDGEMENT

This work is part of the AQUA3S project. AQUA3S has received funding from the European Union's Horizon 2020 research and innovation programme under grant agreement No.832876. Content reflects only the authors' view.

REFERENCES

[1] Garcia, D., Puig, V. & Quevedo, J., 2020. Prognosis of water quality sensors using advanced data analytics: Application to the Barcelona Drinking Water Network. *Sensors*, **20**(5), p. 1342, 2020. DOI: 10.3390/s20051342.

[2] Grbčić, L., Lučin, I., Kranjčević, L. & Družeta, S., A machine learning-based algorithm for water network contamination source localization. *Sensors*, **20**(9), p. 2613, 2020. DOI: 10.3390/s20092613.

[3] Jiang, W., He, G., Long, T., Ni, Y., Liu, H., Peng, Y., Lv, K. & Guizhou Wang, G., Multilayer perceptron neural network for surface water extraction in Landsat 8 OLI satellite images. *Remote Sensing*, **10**(5), p. 755, 2018.

[4] Altenau, E.H., Pavelsky, T.M., Durand, M.T., Yang, X., Prata de Moraes Frasson, R. & Bendezu, L., The Surface Water and Ocean Topography (SWOT) Mission River Database (SWORD): A global river network for satellite data products. *Water Resources Research*, **57**(7), e2021WR030054, 2021.

[5] Johns, R., Community change: Water management through the use of social media, the case of Australia's Murray–Darling Basin. *Public Relations Review*, **40**(5), pp. 865–867, 2014.

[6]   Harou, J.J., Garrone, P.A.O.L.A., Rizzoli, A.E., Maziotis, A., Castelletti, A., Fraternali, P., Novak, J., Wissmann-Alves, R. & Ceschi, P.A., Smart metering, water pricing and social media to stimulate residential water efficiency: Opportunities for the smarth2o project. *Procedia Engineering*, **89**, pp. 1037–1043, 2014.

[7]   AQUA3S. https://aqua3s.eu/. Accessed on: 28 Feb. 2022.

[8]   Smart Data Models. https://smartdatamodels.org/. Accessed on: 28 Feb. 2022.

[9]   The ETSI standard. https://www.etsi.org/deliver/etsi_gs/CIM/001_099/009/01.01.01_60/gs_cim009v010101p.pdf. Accessed on: 28 Feb. 2022.

[10]   Heath, T. & Bizer, C., Linked data: Evolving the web into a global data space. *Synthesis Lectures on the Semantic Web: Theory and Technology*, **1**(1), pp. 1–136, 2011.

[11]   Orion-LD repository. https://github.com/FIWARE/context.Orion-LD. Accessed on: 28 Feb. 2022.

[12]   Cygnus-LD repository. https://github.com/telefonicaid/fiware-cygnus. Accessed on: 28 Feb. 2022.

[13]   Geoserver website. http://geoserver.org/. Accessed on: 28 Feb. 2022.

[14]   WebDAV website. http://www.webdav.org/. Accessed on: 28 Feb. 2022.

[15]   Python version 3.6 website. https://www.python.org/downloads/release/python-360/. Accessed on: 28 Feb. 2022.

[16]   EPANET website. https://www.epa.gov/water-research/epanet. Accessed on 28 Feb. 2022.

CATEGORISING AREA MODELS FOR STORMWATER FEES AT PROPERTY LEVEL: A LITERATURE REVIEW

ULF RYDNINGEN[1], GEIR TORGERSEN[2] & JARLE TOMMY BJERKHOLT[3]
[1]Department of Civil and Environmental Engineering, Norwegian University of Life Sciences, Norway
[2]Department of Engineering, Østfold University College Fredrikstad, Norway
[3]Department of Process, Energy and Environmental Technology, University of South-Eastern Norway, Norway

ABSTRACT

Worldwide, the increasing challenges due to stormwater run-off in urban areas are well known. Authorities need to be prepared for emergency situations and have plans for preventive measures to avoid flooded properties and public grounds. Several studies highlight that homeowner's knowledge and awareness of their own flood risk, will lead to better protection and less damage. What is probably less focused is that preventive-measures within your own property will also help to reduce the flood risk for your neighbours settled at a lower site. Stormwater fee derived from the area model can be seen both as an instrument to motivate property owners to manage rainwater in a more sustainable way, and a way of financing public infrastructure related to stormwater. Many cities and states worldwide have already introduced area models as a basis for calculating stormwater fee at property level. There are many models which range from very simple and rough calculations to more complex and detailed. In some countries, e.g., USA, differentiated stormwater fees have been used for decades, while for example in Norway this is still a controversial topic. In this study, we will conduct a literature review of area models, which aim to describe what a single property should pay in stormwater fee. Which model is best, depends entirely on the goals you want to achieve. Based on the literature review, our understanding is that more attention will be paid on area models if there is a clear connection between instrument and goal. In this article we aim to categorize and group the different models and describe for which goals they are best suited.

Keywords: stormwater fee, barriers, incentives, values, criteria, policy making.

1 INTRODUCTION

Urbanization continues to be the driving force for global growth [1]. The battle for space, which functions to be taken care of within the urban areas, is constantly increasing. At the same time as the urban building density and proportion of impermeable surfaces increases, we face the consequences of climate changes, among others more frequent torrential rainfall, that demands increased local management of stormwater. Large and small cities are struggling on how to deal with this continuously increasing stormwater problems, caused by climate change and decades of urbanization [2]–[8]. In many countries, we observe that local authorities have adopted stormwater fees as a source of revenue to finance maintenance, operation and costly upgrading of stormwater infrastructure [9]–[13]. In the United States, the first municipalities introduced stormwater fees in 1964, and today at least 1,851 local governments in 41 states have this [14]. The emergence of the implementation of stormwater fees across the United States over the past five decades reflects a significant shift in fiscal responsibility for the operation, maintenance, and improvement of public infrastructure systems at the local level [15]. As a result changes in stormwater management policies can lead to intensifying conflicts between urban development and stormwater management [16]. Achieving sustainable urban development requires a balance between economic, social, and environmental assessments in the municipal decision-making processes. To choose good local strategies for sustainable urban development, it is important to understand the barriers that can be encountered in the design and implementation of the desired policies. The aim of this article is to provide an overview of the economic instruments used internationally aiming

WIT Transactions on The Built Environment, Vol 208, © 2022 WIT Press
www.witpress.com, ISSN 1743-3509 (on-line)
doi:10.2495/FRIAR220081

among others to reduce the risk of urban flooding and pollution. For this paper we have identified the following research questions: (1) When introducing stormwater fee; what kind of models at property level to choose between, and how to categorise them? (2) What are the barriers and how to choose model when introducing this fee?

2   CATEGORISATION METHODOLOGY

2.1   Purposes of introducing a stormwater fee

There will be various reasons for implementing stormwater fees, and we will evaluate the purposes described in the literature. It can be (1) to raise money to pay for the running costs, maintenance and the development of new stormwater measures; (2) stormwater fees can be used as incentives to change people's behaviour to manage stormwater on their own plot; (3) in places where there is a shortage of water, water fees can be used to encourage use of stormwater in a more sustainable way, for example for flushing toilets, irrigation of gardens and plants etc.

2.2   Principles on which models for stormwater fees may be based

In determining which considerations are to be prioritized for the choice of economic model for stormwater fees, several trade-offs must be made. The complexity of the models will also reflect the extent to which the different values can be considered. We have chosen to evaluate our findings in the literature according to the considerations and values shown in Table 1.

Table 1: Values on which stormwater fees can be chosen.

Values	Explanation
Simplicity	The calculation of the stormwater fees should be easy to understand for residents and easy to calculate and implement for the local municipality's administration.
Sufficiency	The fees collected must cover both investments, operating and maintenance costs, as well as ensuring that the long-term management of the system is sustainable.
Equity	If equitable revenue responsibility across customer class is important to a local government, moving away from a stormwater fee structure that charges flat fees to all customer classes would be advisable.
Legal	Implementation of the fees is legally justifiable.
Provenance neutrality	The total water and sewage fees should not be changed after the introduction of a separate stormwater fee. But the distribution between the various fees is changing.
Behaviour change	Incentives leading people to change their behaviour e.g. reducing stormwater runoff from their own property.
Polluter pays	In environmental law, the polluter pays principle is enacted to make the party responsible for producing pollution by paying for the damage done to the natural environment.
Precautionary principle	The precautionary principle is an approach to innovations with potential to cause harm when extensive scientific knowledge on the matter is lacking.

Criteria are often associated with values, but they are not synonymous. Criteria may be applied to any number of different levels of experience. Values, on the other hand, are at the same logical level as beliefs. From this perspective, values are similar to what are called "core criteria". Criteria can also be said to be a way of measuring the presence of a value. We have arrived at a conceptualization of values based on three propositions: (a) Values cannot be directly observed; (b) Values engage moral considerations; and (c) Values are conceptions of the desirable [17, p. 28]

2.3 Barrier types

In a comprehensive European study [18], seven types of barriers were identified, these are barriers that affect the ability to transfer policy to action in different ways. Barriers are also studied for spatial planning and policy integration [19], [20] and for climate change adaptation [21].

The barriers overlap to a certain extent and can therefore not be said to be mutually exclusive.

In 1991, economist Douglas North published an article entitled "Institutions" in the *Journal of Economic Perspectives* [22]. North defines institutions as "human-designed constraints that structure political, economic, and social interactions". Restrictions, as described by North, are designed as formal rules (constitutions, laws, property rights) and informal restrictions (sanctions, taboos, customs, traditions, codes of conduct), which usually help maintain order and security within a market or society. Another way of explaining the concept is that institutions are established and stable patterns of behaviour that define, control and limit action. With North's definition of institutions, we choose to classify all the barriers in Table 2, except the technological ones, as institutional barriers. The technological barriers, in this context, can be which parameters are to be included in the models and how accurate data are needed to calculate stormwater fees.

3 METHODS AND DATA

Research literature regarding stormwater fees has been found in various databases, available from the respective universities in which the authors are employed. Keywords in the literature search have been *stormwater fee, stormwater taxation, stormwater funding*. Alternatively, we have also combined the keywords *financing* and *stormwater*. The purpose of this study has also been to investigate barriers and opportunities which also have transfer value for the implementation of stormwater fees. We have also reviewed literature on local stormwater management and searched for articles which deal with *barriers, success factors* and *enablers* in general terms. Our inclusion criteria were peer-reviewed scientific articles from the last twenty years. We have read the articles that met these search criteria to find if they are of interest to our research on various principles for determining stormwater fees and if the articles are mentioning any barriers. After completing searches on these keywords, we have used "chain search" [23, p. 193] where we have studied additional literature mentioned in the articles' reference lists. This means that you find relevant literature in that one text leads to the next. The method has its strength in that it leads from a good reference to another, and in this way, you can follow the development of the arguments through the literature search. The weakness of the method is that you may miss references to other understandings and disagreements in relation to the one you started the chain by. These articles that we find in chain searches, may also be older than the «primary articles», which means that we do not continue with chain searches if the articles are older than twenty years. Our selection criteria

Table 2: Different barrier types and associated challenges.

Barriers	Policy formulation/design	Policy implementation
Cultural	Lack of acceptance and support from the public or affected actors	Lack of support from implementing actors. Low public acceptance
Political	Missing or unstable majority of interests in political choices. Internal tensions in key political parties	Political interference in the implementation phase. No political champion for stormwater fee policy making and implementation
Legal/ regulatory	Lack of or illegitimate legal basis for policy or action design	Missing or illegitimate legal basis for implementation of measures. Developing a new legal framework takes time
Organisational	Unclear and/or conflict-filled division of roles or cooperation between actors.	Unclear placement of responsibilities, lack of capacity or conflict-filled areas of cooperation
Knowledge	Unclear or conflict-filled perception of the connection between measures and goal achievement	Lack of knowledge about methods of implementation. Underestimation of the extent to which a new legal framework would be needed
Economic	Lack of or insufficient funding for political choices or measures	Lack of or insufficient financing of implementation despite formal commitments
Technological	Lack of necessary or mature technology for action	Implemented technology is inadequate or inefficient

also exclude all articles that are not written in English or Scandinavian languages. We have limited this review to articles from the last 20 years, with a few exemptions for articles that are characterised as "reference articles". These are journal papers which are regarded of such great importance that we have included them. With these subjective inclusion and exclusion criteria, we have reduced the number of relevant articles for this literature study to 113. This literature has been carefully reviewed to identify if they are relevant to our research, and finally we ended with 56 references.

4 RESULTS

4.1 Calculation basis for stormwater fees found in the literature

The equivalent residential housing unit [9]–[11], [13]–[16], [24]–[35] approach, uses the average impervious area of a property as a standard unit to determine the stormwater fee. Residential equivalent factor [10], [14], [29] calculates the average runoff volume for a selected stormwater event for properties with the same zone status (e.g., single dwelling), and all properties in this zone-category are charged the same fee. For the gross property area, a stormwater fee is imposed on the property's gross area. This model assumes that the entire property contributes to run-off or that all properties are equally developed. The distributed transportation alternative [10], [13] considers the stormwater management of municipal roads and calculates the toll based on the average mileage for a specific user. The hydrologic alternative fee [10], [11], [36] is based on the characteristics of the individual property, such as soil type, topography, impervious area, and requires detailed information about each plot. Incentive scheme is used to encourage residents to manage as much stormwater as possible on their own property [9], [12], [13], [32], [37]. To improve runoff quality [9], [25], to reduce the demand on the sewer infrastructure [37], to install LIDs [4], [9], [26] a discount scheme can be built in to compensate for this. In intensity of development [9], [14], [27], [35], [38] stormwater fee is usually very similar to the coefficient of runoff [35]. The gross area and the intensity of development methods assign factors for different land use types [31]. For the tier fee [29] fees are charged by classifying property plots in categories based on the rate of impervious area, land use purposes, etc. The flat fee is a funding mechanisms that charge a flat rate to users of a stormwater conveyance system [24], [29] and is based on the size of a property, and on the average stormwater burden their property type contributes [32]. Fee types that are easy to administer (e.g., flat fees) is not fully representing the stormwater contribution from the parcels [29]. The dual fee [24], [29] calculation method divides properties into residential or non-residential and burdens the classifications differently. For the water usage fee [29], [32] the fee level is according to the household water usage. Runoff volume and rates can be determined using a number of factors (usually the impervious area) [13]. The parcel area rates is based on the size of parcel [29]. The cap and trade [9], [15] evaluation is based on criteria quota trading with discharge limits for stormwater runoff. The municipality sets a limit for the discharge of stormwater on the main sewer system. The maximum permitted discharge is then distributed among residents. Quota trading is encouraged, where quotas give permission to release a certain amount of stormwater at the property level. For offset program [25] reimbursement of overpaid income tax that the landowner has paid in is set off against debt due for stormwater fee. The stormwater fee with fixed and variable component is a proposed method for taxation based on individual parcel assessment, described by Barton [36], Peterson et al. [39] and Godyń et al. [40]. For these models a fee specific for the property can be calculated through the combination of detailed land use maps, a hydrological model, estimates of current and expected costs of stormwater networks and treatment. For the Pigouvian instrument tax concept [12], [25], [33] one assumes that optimal tax on pollution should be a direct tax that corresponds to the marginal external damages caused by the pollution. Traditional "command-and-control" regulations set uniform standards for all sources, with the most common being technology- and performance-based standard. Command-and-control regulation sets specific limits for pollution emissions and/or mandates that specific pollution-control technologies that must be used [41]. Ad valorem tax [16], [21] is a tax whose amount is based on the value of a transaction of property. Funds are not dedicated and stormwater programs must compete with other programs for funding [32]. The parcel area fee is based on size of parcel.

Even though several of the articles reviewed discuss implementation challenges, we rarely find any discussions about which values, or barriers formed the basis for the model. Table 3 shows the evaluation and the criteria that seem to be emphasized for the different models. Some models are very common, which we interpret as being easy to understand for residents, they are politically acceptable and easy to administer. Such simple models, on the other hand, are not quite fair or sufficient, as they have simplified calculations of the individual parcels in terms of infiltration properties and stormwater load.

4.2 How to overcome the implementation barriers in a stormwater context

When planning to introduce a fee for local stormwater management, barriers among decision-makers, residents [52], [53] or the local administration may occur. Ngyen et al. [33, p. 157]

Table 3:   Evaluation of stormwater fee models.

| Fee models | Purpose of the stormwater fee models | | | | | Criteria for calculation | | | | |
| | Economic goals | | Environmental goals | | | Property | | | Hydrology/ stormw. vol. | |
	Financing	Incentive – reduced fee	Reuse stormwater	Prevent floods	Prevent pollution	Total area	Impervious area	Stormwater runoff	SUDS measures	Terrain slope/ hydrolog. proc.
1 Equivalent residential unit [10], [11], [13], [14], [26], [27], [29], [38], [42]	x			x		x	x	x		
2 Residential equivalent factor [10], [14], [29]	x			x		x	x	x		x
3 Gross property area [31], [35]	x			x		x				
4 Distributed transportation alternative [10], [13]	x			x				x		
5 Hydrologic alternative [10], [11], [36]	x	x		x		x	x	x		x
6 Incentive scheme [4], [9]–[13], [25], [32], [35], [37], [43], [44]	x	x			x				x	
7 Intensity of development [31], [35]	x			x		(x)	(x)			
8 Tier fee [24], [29]	x			x		x	x			
9 Flat fee [11], [24], [28], [29], [31], [32], [39], [45]	x			x		x		(x)		
10 Dual fee [24], [29]	x			x		x	x			
11 Water usage fee [13], [24], [29], [31], [36]	x	x	x	x						
12 Cap and trade [9], [15], [46]	x	x		x				(x)		
13 Offset program [4], [9], [12], [29], [34], [47], [48]	x	x	x	x						
14 Pigouvian instrument [9], [11], [15]	x	x	x	x	x					
15 Command and control [9], [15], [41], [46], [49], [50]	x	x		x				x		
16 Ad valorem tax [27], [32]	x			x		(x)				
17 Parcel area [11], [12], [24], [29]	x			x		x				
18 Fixed and variable component [32], [36], [40], [51]	x	x	x	x	x	x	x	x	x	x

claims that "the close cooperation of various levels of government administration is a vital factor for successful implementation". Many of the barriers are difficult to overcome because they are systemic and embedded within organizational cultures, practices and processes [54]. Griggs writes "How programs are financed, depends on their scales of operation, how they are organized, and on the community's approach to taxation and use of fees. In a state with the enabling legislation and successful utilities in place, it makes sense to adopt stormwater fees if they are politically acceptable" [12].

Key *success factors* in implementing new policies and measures are described by Åkerman et al. [18]: *Cultural barriers*: Carefully prepare information and communication strategies. *Political barriers:* Having a project champion who is backing up the stormwater fee policy and manages to create legitimacy. A political leadership developing over time, with a network of policy makers and experts who is working intensively for successful policy implementation. *Legal barriers*: Having an existing legal framework that supports this kind of initiative. *Organisational barriers*: A clear pioneering spirit among key professionals involved in the preparations. Existing institutional framework with clear mandates, roles and responsibilities. *Knowledge barriers*: Preparatory work and expert knowledge exist and is easy to access and bring into the process. *Economic barriers*: Acceptable initial capital costs and financing. *Technological barriers:* Available technology that work well.

5 DISCUSSION

We have managed to identify eighteen principles for fee or tax determination in the reviewed literature.

When national laws refer to stormwater only as a pollution problem, the law is then probably based on the precautionary principle. If laws must be changed to allow the introduction of stormwater fees, i.e., stormwater is regarded as a quantity problem, then the law must be founded on the polluter pays principle. When municipalities emphasize the polluter pays principle, the calculation of the stormwater fee should ideally include the costs the stormwater run-off from each individual property potentially contributes to the surroundings. Furthermore, it should also include the costs this runoff will impose on the local authority working to prevent damages. The most common way of calculating the stormwater fee is to apply the equivalent residential unit model (ERU) where the fee is determined based on the proportion of permeable surfaces on each property. Basically, this is a fee for stormwater that occurs from sealed surfaces on the site, such as roofs, driveways etc. and from which the stormwater is led to the public stormwater system. In practice, it is hard to design a fee which is regarded as fair for all types of properties. To reduce the administrative costs and to make the fee easier to deal with, a fee which is scaled proportionally to the plot area has therefore been introduced. In Växjö in Sweden, the design of the fee is currently divided into three categories (detached houses, other buildings and plots without buildings) [44]. This could possibly challenge some landowners' view on what can be accepted as fair.

Some municipalities have enacted the principle that despite the introduction of a new stormwater fee, the total amount of charges should remain unchanged.

Our research reveals that all the previously mentioned barriers are important for local municipalities to have in mind when implementing storm water fee at a local level. Which barriers that are the major obstacles during the process will vary a lot and requires detailed knowledge of local conditions. The organization of the water sector can be fragmented, in that various administrative units have shared responsibility for e.g., stormwater management from the built environment. For example, in the literature we have found that in some cities the Road Department is responsible for the stormwater management [10], while in other cities this belongs to the Water and Sewerage Department. In addition to organizational barriers, we also find examples of financial barriers. This might be that the agencies responsible for stormwater management operate with insufficient budgets. To overcome these challenges the legislation must be changed so it provides the right to collect stormwater fees sufficient to cover the full cost of producing the services. For some landowners, the burden of a high stormwater fee may motivate them to set up preventive measures that can reduce the stormwater fee. This can be achieved by building more green infrastructure or reduce non-permeable surfaces on their individual property [29]. This can then lead to some unintended consequences for the municipality. If high number of landowners install measures that infiltrate, delay and/or detain stormwater on their own property, and thus pay a reduced stormwater fee, then the monthly income of the municipality can be reduced so that the municipality must increase the fee rates for stormwater management [29]. This possibility of fee reduction gives local politicians another dilemma; what about those properties that do not have the opportunity to install any green stormwater measures? In dense urban areas, there will be little space available for new stormwater measures, except green roofs and walls on some buildings. At the same time, when much of the older urban development was approved for construction permits in earlier times, it was not relevant to set any additional requirements for stormwater management at the individual properties. It may seem unfair for these landowners to be imposed a maximum stormwater fee, since it is not possible to have mitigating measures on their own property. An additional dilemma is related to the fact that the types of fees that are easy to administer will not fully represent the stormwater contribution from the individual properties. A consequence of introducing a model that more accurately represents the stormwater load from each individual property, will inevitably be more administratively demanding, and contain a larger number of factors that might be difficult or costly to obtain. Stormwater fees have not only been introduced to finance necessary measures, or to cover costs. Using low impact development (LID) approach to site development and stormwater management, the basic principle is to use nature for better source control. This is accomplished through sequenced implementation of runoff prevention strategies, runoff mitigation strategies, and finally, treatment controls to remove pollutants. Wilkerson et al. [55] mentions "awareness barriers" and explains this by saying that LID is a relatively new practice that most households are unaware of. At the same time, it turns out that achieving a voluntary development of LID on existing city properties only through the use of information campaigns has little effect. The incentive scheme is not intended to only provide funding for necessary stormwater measures, but it should also lead to actions to reduce runoff. The use of incentives may give more attention to the topic, but at the same time the fee level must be of a certain size to lead to a change in behaviour.

In practice, some models discussed here are little in use, such as The cap and trade and hydrologic alternatives. The latter is based on characteristics associated with the individual property such as soil type, topography, impermeable surfaces, and requires very detailed information [11], [36]. The purpose of using a stormwater fee may vary. About a third of the EU's territory is exposed to permanent or temporary water shortages. Thus, more sustainable water management can prevent water shortages. Rain water harvesting system can reduce water shortage during dry periods in arid regions, and a fee model that is linked to "water usage" may then be relevant [56]. Water consumption (and the fee level) can then be reduced by collecting rainwater and using it for example to cleaning, watering plants, and flushing toilets.

The Pigou instrument is a fee model based on payment for externalities. These are effects that you inflict on others that you do not include in your choices. An example that can be used, is when a property at the upper part of a catchment area cause flooding of the properties below. The landowner who resides at the top, neither knows that his actions affect others, nor knows that stormwater from his property causes problems for others who live further down. Economic theory defines this as a negative externality. Pigou's solution to the problem is to increase the private marginal cost by introducing a tax on pollution or inconvenience that the individual inflicts on others. At the same time, the environmental tax contributes to fulfilling a principle that polluters must pay, and the tax contributes to changing production and consumption patterns over time. Ad valorem tax is typically imposed at the time of a transaction, as in the case of a sales tax or value-added tax (VAT). Funds are not dedicated, and thus can be insufficient or erratic; stormwater programs must compete with other programs for funding. The taxpayer payments will in no way reflect stormwater burden, thus there is no incentive to modify actions [32].

When a municipality is planning to introduce a differentiated fee model, stakeholders should be involved through participation processes to identify values and barriers. Technical, economic, and environmental aspects as well as local knowledge could then be included in the planning processes at an early stage.

6 CONCLUSION

In this study, our focus has been to identify differences between calculation models of stormwater fees. To answer the research questions, we studied how stormwater fees were calculated in eighteen countries. We have identified 18 different models, that describes the calculation of stormwater fees. Some of these models are simple and widely used and therefore assumed accepted by both the population and local politicians as an appropriate tool for calculating stormwater fees. More sophisticated models that requires more detailed data about each property can be regarded as fairer but are more demanding to install and run. The main intention of this study was to identify fee models developed on property level and barriers associated with the implementation of such models. However, this study also included some fee models that are based solely on economic principles and not on characteristics on individual property. When implementing stormwater fee there exists both technical and institutional barriers. For successful implementation these barriers are important to identify at an early stage.

In recent years, stormwater management has evolved from a site-specific, technical issue mainly handled by engineers and water professionals to be an interdisciplinary field including engineers, landscape architects, urban planners, and citizens. This entails decision-making across different norms, values and work-practices that sometimes be challenging.

When implementing stormwater fee, the fee must be comprehensible and fair to citizens. But at the same time, it must be reliable and easy to set up. Essential basis data for calculating the fee, e.g., property conditions, precipitation and investment costs should be simple to collect and maintain for municipalities. The model must also be possible to manage for smaller municipalities with limited time, human resources, and expertise.

If local authorities emphasize a model that can be regarded as fair and simple, a stormwater fee structure where the property owners pay a fee proportional to the plot area, should be chosen. To motivate citizens to reduce or delay rainwater from their own property, permeable surfaces, ponds etc. should be included in the model. The ERU-model can then be the starting point for developing a fee structure. Although the intention is to create the model as fair as possible, it will soon become difficult for the citizens to understand as well as complicated for professionals to maintain. If the main goal for the fee is to generate revenue to cover running costs and maintenance of the storm water system based on a simple model, flat fee is preferable. Furthermore, tier fee is an option if you still want to keep the model simple, but at the same time differentiate the payment from properties to some extent.

WIT Transactions on The Built Environment, Vol 208, © 2022 WIT Press
www.witpress.com, ISSN 1743-3509 (on-line)

In preparation for the introduction of stormwater fee, local politicians, municipal staff, and landowners should be involved at an early stage in order to map which values and barriers are important to emphasize when designing the fee model.

ACKNOWLEDGEMENTS

We would like to extend thank our colleagues Ass. Prof. Vegard Nilsen and Ass. Prof. Kim Paus for valuable comments and input during the writing process.

REFERENCES

[1]   UN-Habitat, *World Cities Report 2020: The Value of Sustainable Urbanization*, United Nations Human Settlement Programmen (UN-Habitat): Nairobi, Kenya, 2020.
[2]   Abebe, Y., Adey, B. & Tesfamariam, S., Sustainable funding strategies for stormwater infrastructure management: A system dynamics model. *Sustainable Cities and Society*, **64**, 102485, 2021.
[3]   Boguniewicz-Zabłocka, J. & Capodaglio, A.G., Analysis of alternatives for sustainable stormwater management in small developments of Polish urban catchments. *Sustainability (Basel, Switzerland)*, **12**, 10189, 2020.
[4]   Carlson, C. et al., Storm water management as a public good provision problem: Survey to understand perspectives of low-impact development for urban storm water management practices under climate change. *Journal of Water Resources Planning and Management*, **141**(6), 04014080, 2015.
[5]   Jotte, L., Raspati, G. & Azrague, K., *Review of Stormwater Management Practices*, Klima 2050. SINTEF Building and Infrastructure, 2017.
[6]   Lu, G. & Wang, L., An integrated framework of green stormwater infrastructure planning: A review. *Sustainability*, **13**(24), 2021.
[7]   Van Oijstaeijen, W., Van Passel, S. & Cools, J., Urban green infrastructure: A review on valuation toolkits from an urban planning perspective. *J. Environ. Manage.*, **267**, 110603, 2020.
[8]   Veiga, M.M., Castiglia-Feitosa, R. & Marques, R.C., Analyzing barriers for stormwater management utilities. *Water Supply*, **21**(4), pp. 1506–1513, 2021.
[9]   Parikh, P. et al., Application of market mechanisms and incentives to reduce stormwater runoff. *Environmental Science and Policy*, **8**(2), pp. 133–144, 2005.
[10] Fisher-Jeffes, L. & Armitage, N.P., Charging for stormwater in South Africa. *Water S. A.*, **39**(3), 2013.
[11] Tasca, F.A., Assunção, L.B. & Finotti, A.R., International experiences in stormwater fee. *Water Sci. Technol.*, **2017**(1), pp. 287–299, 2017.
[12] Grigg, N.S., Stormwater programs: Organization, finance, and prospects. *Public Works Management and Policy*, **18**(1), pp. 5–22, 2013.
[13] Tasca, F.A., Finotti, A.R. & Goerl, R.F., A stormwater user fee model for operations and maintenance in small cities. *Water Sci. Technol.*, **79**(2), pp. 278–290, 2019.
[14] Campbell, W. & Bradshaw, J., *Western Kentucky University Stormwater Utility Survey 2021*, Western Kentucky University: Bowling Green, Kentucky, p. 80, 2021.
[15] Chalfant, B., Resistance is (sometimes) futile: The diffusion and (occasional) demise of stormwater fees in the United States. *Proceedings of the Water Environment Federation*, **2018**, pp. 1015–1036, 2018.
[16] Bertram, N.P. et al., Synergistic benefits between stormwater management measures and a new pricing system for stormwater in the city of Hamburg. *Water Science and Technology*, **76**(6), pp. 1523–1534, 2017.

[17] Deth, J.W.v. & Scarbrough, E., *The Impact of Values*, Oxford University Press: Oxford and New York, 1995.

[18] Åkerman, J. et al., How to manage barriers to formation and implementation of policy packages in transport. *OPTIC. Optimal policies for transport in combination – 7th framework programme: Theme 7 transport*, p. 73, 2011.

[19] Stead, D. & Meijers, E., Spatial planning and policy integration: Concepts, facilitators and inhibitors. *Planning Theory and Practice*, **10**(3), pp. 317–332, 2009.

[20] Oukes, C., Leendertse, W. & Arts, J., Enhancing the use of flood resilient spatial planning in Dutch water management: A Study of barriers and opportunities in practice. *Planning Theory and Practice*, pp. 1–21, 2022.

[21] Oulahen, G. et al., Barriers and drivers of planning for climate change adaptation across three levels of government in Canada. *Planning Theory and Practice*, **19**(3), pp. 405–421, 2018.

[22] North, D.C., Institutions. *Journal of Economic Perspectives*, **5**(1), pp. 97–112, 1991.

[23] Rienecker, L. et al., *Den gode oppgaven: håndbok i oppgaveskriving på universitet og høyskole*. Fagbokforl.: Bergen, 2006.

[24] Aladesote, O., Stormwater management utility fees: A review. *International Journal of Research Publications*, **40**(1), pp. 1–13, 2019.

[25] Chouli, E., Aftias, E. & Deutsch, J.-C., Applying storm water management in Greek cities: Learning from the European experience. *Desalination*, **210**, pp. 61–68, 2007.

[26] Cousins, J.J. & Hill, D.T., Green infrastructure, stormwater, and the financialization of municipal environmental governance. *Journal of Environmental Policy and Planning*, **23**(5), pp. 581–598, 2021.

[27] Cyre, H.J., Stormwater management financing. *International Public Works Congress*, American Public Works Association: Houston, Texas, 1982.

[28] Environmental Commissioner of Ontario, *Urban Stormwater Fees: How to Pay for What we Need*, Environmental Commissioner of Toronto: Ontario, Canada, p. 34, 2016.

[29] Fedorchak, A., Dymond, R. & Campbell, W., The financial impact of different stormwater fee types: A case study of two municipalities in Virginia. *Journal of the American Water Resources Association*, **53**(6), pp. 1483–1494, 2017.

[30] Grigg, N., Is a stormwater fee a rain tax? *Journal of Legal Affairs and Dispute Resolution in Engineering and Construction*, **11**, 04519017, 2019.

[31] Kea, K., Dymond, R. & Campbell, W., An analysis of patterns and trends in United States stormwater utility systems. *JAWRA Journal of the American Water Resources Association*, **52**(6), pp. 1433–1449, 2016.

[32] Keeley, M., Using individual parcel assessments to improve stormwater management. *Journal of the American Planning Association*, **73**(2), pp. 149–160, 2007.

[33] Nguyen, T.T. et al., Implementation of a specific urban water management: Sponge City. *Science of The Total Environment*, **652**, pp. 147–162, 2019.

[34] Roy, A.H. et al., Impediments and solutions to sustainable, watershed-scale urban stormwater management: Lessons from Australia and the United States. *Environ. Manage.*, **42**(2), pp. 344–359, 2008.

[35] Tucker, S. et al., *Guidance for Municipal Stormwater Funding*. epa.gov, p. 140, 2006.

[36] Barton, D.N., Overvannsgebyr: internasjonal erfaring og muligheter i Norge. *Fagtreff i Norsk Vannforening: Nytt vann i gamle byer – Forskning og nye muligheter*, Streaming video: NINA, 2021.

[37]   Brears, R.C. (ed.), Copenhagen becoming a blue-green city. *Blue and Green Cities: The Role of Blue-Green Infrastructure in Managing Urban Water Resources*, Palgrave Macmillan UK: London, pp. 99–126, 2018.

[38]   Cyre, H., Developing and implementing a stormwater management utility: Key feasibility issues. *International Public Works Congress*, American Public Works Association: New Orelans, Louisiana, p. 7, 1986.

[39]   Peterson, K. et al., A Review of funding mechanisms for US floodplain buyouts. *Sustainability (Basel, Switzerland)*, **12**, 10112, 2020.

[40]   Godyń, I., Muszyński, K. & Grela, A., Assessment of the impact of loss-of-retention fees on green infrastructure investments. *Water (Switzerland)*, **14**(4), 2022.

[41]   Schoeman, J., Allan, C. & Finlayson, C.M., A new paradigm for water? A comparative review of integrated, adaptive and ecosystem-based water management in the Anthropocene. *International Journal of Water Resources Development*, **30**(3), pp. 377–390, 2014.

[42]   Olstad, F. & Thuve, M., *Vurdering av finansieringsmodeller for overvann. Rapportutkast*, BDO: Oslo, p. 52, 2015.

[43]   Milon, J.W., The polluter pays principle and Everglades restoration. *Journal of Environmental Studies and Sciences*, **9**(1), pp. 67–81, 2019.

[44]   Paus, K.H. et al., *Overvannsarbeid i utlandet. Virkemidler for å redusere nedbørbetinget oversvømmelse i urbane områder*, COWI: Oslo, p. 43, 2015.

[45]   Cameron, J. et al., User pay financing of stormwater management: A case-study in Ottawa-Carleton, Ontario. *Journal of Environmental Management*, **57**(4), pp. 253–265, 1999.

[46]   Petersen, B. & Ducos, H. , Justice in climate action planning. *Strategies for Sustainability*, eds R. Lozano & A. Carpenter, Springer: Cham, Switzerland, p. 310, 2022.

[47]   Hirschman, D. & Battiata, J., *Urban Stormwater Management: Evolution of Process and Technology*, Springer International Publishing, pp. 83–120, 2016.

[48]   Niu, H. et al., Scaling of economic benefits from green roof implementation in Washington, DC. *Environ. Sci. Technol.*, **44**(11), pp. 4302–4308, 2010.

[49]   Dhakal, K.P. & Chevalier, L.R., Urban stormwater governance: The need for a paradigm shift. *Environmental Management*, **57**(5), pp. 1112–1124, 2016.

[50]   Grigg, N.S. (ed.), *Purposes and Systems of Water Management*, in *Integrated Water Resource Management: An Interdisciplinary Approach*, Palgrave Macmillan UK: London, pp. 33–66, 2016.

[51]   Barton, D.N. et al., Brukerfinansiert klimaberedskap? En beregningsmodell for overvannsgebyr i Oslo. *VANN*, **56**(04), pp. 341–358, 2021.

[52]   Pierce, G. et al., Environmental attitudes and knowledge: Do they matter for support and investment in local stormwater infrastructure? *Society and Natural Resources*, **34**(7), pp. 885–905, 2021.

[53]   Kollmuss, A. & Agyeman, J., Mind the gap: Why do people act environmentally and what are the barriers to pro-environmental behavior? *Environmental Education Research*, **8**(3), pp. 239–260, 2002.

[54]   O'Donnell, E.C., Lamond, J.E. & Thorne, C.R., Recognising barriers to implementation of blue-green infrastructure: a Newcastle case study. *Urban Water Journal*, **14**(9), pp. 964–971, 2017.

[55]   Wilkerson, B., Romanenko, E. & Barton, D., Modeling reverse auction-based subsidies and stormwater fee policies for low impact development (LID) adoption: A system dynamics analysis. *Sustainable Cities and Society*, **79**, 2021.

[56] Musz-Pomorska, A., Widomski, M.K. & Gołębiowska, J., Financial sustainability of selected rain water harvesting systems for single-family house under conditions of eastern Poland. *Sustainability (Basel, Switzerland)*, **12**(12), p. 4853, 2020.

[57] Kvamsås, H., Addressing the adaptive challenges of alternative stormwater planning. *Journal of Environmental Policy and Planning*, **23**(6), pp. 809–821, 2021.

WATER DEMAND SCALING LAWS AND SELF-SIMILARITY PROPERTIES OF WATER DISTRIBUTION NETWORKS

MANUELA MORETTI, ROBERTO GUERCIO & ROBERTO MAGINI
Department of Civil, Building and Environmental Engineering, Sapienza University of Rome, Italy

ABSTRACT
The design of water distribution networks (WDNs) usually considers deterministic values of nodal water demand, calculated by multiplying the average water demand by an appropriate demand factor, which is the same for all nodes. Obviously, changes in the demand factor produce different, yet perfectly correlated, demand scenarios. Today's large availability of high-frequency water consumption monitoring allows describing water demand in statistical terms. The traditional deterministic approach, characterized by a perfect correlation between nodal demands, leads to an analytical dependency between the hydraulic heads in each of the nodes and the total flow entering the network. On the other hand, if we consider that the nodal demand is described by marginal probability distributions, differently correlated with each other, this result is still valid, but only for the mean. In this work, several scenarios have been generated through stratified random sampling (Latin hypercube sampling). The nodal water demand is described by Gamma probability distributions whose parameters are related to the type and number of users according to suitable scaling laws, derived from historical data sets. The results were obtained considering different types of users and different network topologies and highlighted the possibility of evaluating the mean function of the nodal hydraulic head vs the total entering flow based on the direct acyclic graph (DAG) of the network. Moreover, the dispersion of the data around the mean function was found to be dependent on the properties of the network: dimension and topological structure.
Keywords: scaling laws, water demand scenarios, self-similarity.

1 INTRODUCTION

water distribution network (WDN) sizing, calibration, and management largely depend on the water demand scenarios considered. When dealing with one of these issues, nodal demands are often not known beforehand: they must be measured or estimated. Measuring all nodal demands in an existing network is an unaffordable procedure, and not possible in the design phase.

Traditionally, deterministic water demand scenarios are estimated from annual average consumption measurements. These scenarios are properly scaled by a demand factor, depending on the size and socio-economic characteristics of the area covered by the WDN. Since they derive from a single multiplicative scaling factor, all the scenarios obtained from different demand factors are perfectly correlated with each other.

Different methods for generating stochastic demand scenarios have been proposed in the literature [1], [2], in order to overcome the conceptual limitations of this approach. In particular, the one proposed by Magini et al. [1], uses the observation of the scaling properties of the statistical moments of the demand probability distributions, deriving from the spatial aggregation of water users.

The hydraulic response of the network to a demand scenario is characterized by its values of pipe flow rates and nodal pressure heads. This highlights the mutual interaction between demand, network topology, and characteristics of the pipes. Consequently, this interaction allows predicting pressure at each node under given demand conditions. This allows

WIT Transactions on The Built Environment, Vol 208, © 2022 WIT Press
www.witpress.com, ISSN 1743-3509 (on-line)
doi:10.2495/FRIAR220091

envisaging possible critical issues in some parts of a WDN and accelerating numerical procedures in its hydraulic simulation.

This work presents some analytical evidence regarding the possibility to identifying the hydraulic behaviour of the WDN in stochastically perturbed scenarios with respect to an average demand scenario. In particular, the average scenario results to be adequate for the hydraulic description of the WDN. Finally, to verify the proposed approach, we propose an example on a WDN case from the literature, using synthetically generated stochastic demand scenarios.

2 DEMAND SCENARIOS

The total flow entering the network and, even more, nodal water demand, represent the most uncertain input parameters in the hydraulic modelling of WDNs. The estimation of these input parameters differs according to three different conditions: (1) a water-meter dataset is available, (2) a dataset measured from a similar system is used, or (3) no measured datasets are available. In the last case, the annual average demand scenario is usually considered [3], [4].

From this basic scenario, it is also possible to obtain peak demand scenarios, which are useful to determine production and distribution capacity and customers' metering in an existing network. Peak flow is a key factor for this purpose. In the absence of data, the usual approach to estimate the peak flow is to consider a peaking factor (PF). The PF is the ratio between the maximum water demand recorded in a given time interval and the average annual water demand. Increasing the average nodal flow rates by the same factor is equivalent to considering the common behaviours of network users and therefore the perfect correlation of their water demands. This approach does not allow the identification of the probabilistic dispersion of the nodal head that is consequent to the probabilistic dispersion of the water demand in the different scenarios.

However, internationally the use of the average daily water demand (ADD) and the PF is the most widespread method in professional practice. For example, to harmonize the European water supply standards, a peak flow probability method – originally used in Switzerland – was incorporated into EN 806-3. The American counties also propose this methodology, each adopting different peak factors [5]–[7].

However, many recommendations in current design standards have been carried over from previous standards unquestioned and without revision. Furthermore, since the genesis of these measures is not well documented or understood, disagreement between regional methodologies increased.

In this paper, we also want to highlight that no scientific evidence confirms that the annual average scenario, even when appropriately amplified through the PF factor, is the most precautionary scenario for the WDN.

2.1 Stochastic demand scenarios

It's worth pointing out that nodal water demand can greatly influence the model accuracy [8]–[10]. The deterministic approach, despite (and due to) its simplicity of use, does not allow considering the stochastic component of demand. In fact, water demand consists of a deterministic component – linked to the household plumbing fixtures and appliances – and a stochastic component linked to the unpredictable behaviour of users. It is necessary that the cross-correlation between users' water demands is considered in the stochastic component, in order to estimate the real peak scenario. Therefore, real-time estimation of nodal water demand is a major task for the real-time modelling of WDNs. Thanks to the development of smart metering technology, a large amount of measured data has gradually become available in recent years.

The observation of the scaling properties related to spatial and temporal aggregation of water demand measures allowed defining a method for the generation of stochastic demand scenarios [1].

For this purpose, it is assumed that the statistical moments of nodal demand (mean, variance, and cross-correlation) depend on type and number of users according to suitable scaling laws [11]. The development of the scaling laws assumes that the demand can be described by a homogeneous and stationary process, which implies that the aggregated users are of the same type (residential, commercial, industrial, etc.), and that the statistical properties of demand are constant over time.

The water demand statistical features of the single users are obtained from historical datasets and, together with the number of users in each node, represent the main input data for scenario generation. The most appropriate probability distribution to represent the nodal marginals may depend on the number of aggregated users and derives from historical data. In the literature, the most used probability distributions to describe the variability of the demand for aggregated users are the Log-normal, Gamma, and Weibull.

The complete procedure for scenario generation adopted in the application is detailed in Magini et al. [1]. It is based on sampling from the marginal distributions using the Latin hypercube sampling (LHS) [12]. In order to respect cross-correlation between nodal demands a combination of the NORTA model [13], and the Iman–Conover method [14] is used. The LHS is a "stratified sampling" technique that produces a better description of the input probability distribution with fewer iterations compared with a simple random sampling. The NORTA model is a two-step process, first transforming a multivariate normal vector Z into a multivariate uniform vector U, then transforming the latter into the desired input vector. The joint distribution of U is a copula, and any joint distribution can be represented as a transformation of a copula. To improve compliance with the network-demand correlation structure, the restricted pairing Iman–Conover technique is applied to NORTA results. It induces rank correlation by shuffling finite-size samples obtained from NORTA. The appropriate shuffling is determined by ranking the input samples the same as in a reference sample with the desired rank correlation. The demand scenarios obtained represent an improvement of the commonly used deterministic ones. In fact, the proposed model can generate scenarios in which, given the peculiar characteristics of the users, nodal water demand contains the stochastic component. However, empirical analyses of the generated scenarios show that the stochastic component is small, compared to the values assumed by the average demand. These nodal water demand distributions can be defined as "stochastic scenarios".

3 THE WATER DISTRIBUTION NETWORK

A WDN is mainly composed by the pipes transporting water from an inlet with known hydraulic head towards the demand nodes. A looped WDN can be represented by a directed graph that schematizes its geometric structure, planimetrically and altimetrically, and the capacitive characteristics of pipes, i.e., diameters, roughness, and lengths. In accordance with the characteristics of the considered WDN, each water demand scenario gives rise to water flows in pipes along definite directions, which define the orientation of the arcs of the graph.

For any scenario and size of the network, the corresponding oriented graph has no loops, as the hydraulic system is governed by the variational principle of minimum energy dissipation [15] and respects the uniqueness property of the solution [16]. A structure with these properties is called a direct acyclic graph (DAG).

Based on these assumptions, it can be stated that each WDN is characterized by its own basic DAG, which is linked to the average demand scenario.

The direction of the flows in the pipes varies for each scenario that occurs on a network, therefore for each scenario a specific DAG is defined, which can be known solving the hydraulic model of the WDN.

3.1 Water distribution network solvers

Numerous algorithms have been developed over time for solving the mixed set of linear and nonlinear equations governing the steady-state hydraulics of looped WDNs. The different approaches can be divided into local approaches (e.g., the Hardy Cross method [17]), which deal with one equation at a time, and global approaches, which solve simultaneously all the equations. For the second group, it is possible to make use of the Newton–Raphson (NR) linearization method or the linear theory (LT) successive approximation method to treat the system's nonlinear equations. One of the most important resolution methods for looped networks is the LT introduced by Wood and Charles [18]. Nevertheless, after more than thirty years, in the technical and scientific fields, the reference standard for the resolution of hydraulic networks is still the method introduced by Todini and Pilati [19], Todini [20] and Todini and Rossman [21]. Whichever the numerical solution is chosen, the nodal demand scenario represents a critical parameter, from which the response of the network drives.

3.2 The behaviour of WDNs in deterministic water demand scenarios

Given a network with l loops, p pipes and n demand nodes, without tanks inside, the set of resolutive equations can be expressed for the generic node i as follows:

$$Q_{tot} = \sum_{i=1}^{n} q_i,\tag{1}$$

$$\sum_{k=1}^{p} a_{i,j} Q_{k_{i,j}} + q_i = 0, \qquad i = 1,2,\dots,n,\tag{2}$$

$$\sum_{k=1}^{p} \beta_{j,k}\, r_k\, |Q_k|^{\alpha-1} Q_k = 0, \qquad j = 1,2,\dots,l.\tag{3}$$

This system of equations contains as unknowns only the pipe flow rates Q_k and it is determined. In the case of networks served by one or more reservoirs having the same hydraulic head H_0, the value of the hydraulic head H_i at each node i can be evaluated as follows:

$$H_i = H_0 - dH_i = \sum_{k=1}^{p} \gamma_{D_k}\, L_k\, Q_k |Q_k|^{\alpha-1}, \qquad i = 1,2,\dots,n.\tag{4}$$

Given a water demand scenario $D = (q_1, q_2, \dots, q_n)$, the total inflow Q_{tot} is automatically known, so it is possible to define the attenuation coefficient $ac_k = Q_k/Q_{tot}$ of the pipe flows and to link the head losses dH_i in eqn (4) to the total flow entering the WDN:

$$dH_i = \sum_{k=1}^{p} \gamma_{D_k} L_k \left(\frac{Q_k}{Q_{tot}} \cdot Q_{tot} \right)^{\alpha} = \sum_{k=1}^{p} \gamma_{D_k} L_k \left(ac_k \cdot Q_{tot} \right)^{\alpha}. \tag{5}$$

In a looped network, given a demand scenario D, the flow follows many different paths to reach the demand node along which the head loss dH_i is the same. For this reason, the parameters related to the flow rate can be enclosed by a single invariant parameter k at each demand node in the various scenarios:

$$dH_i = k_i \cdot Q_{tot}{}^{\alpha}. \tag{6}$$

As already mentioned, the reference scenario is the average scenario (subscript m), and in a deterministic approach all the other possible scenarios (subscript s) can be obtained by multiplying the former by a factor f. It follows that:

$$Q_{k,s} = f \cdot Q_{k,m}, \tag{7}$$

$$Q_{tot,s} = f \cdot Q_{tot,m}. \tag{8}$$

As a consequence, the head loss is scaled according to a power law:

$$dH_{i,s} = k_i \cdot f^{\alpha} \cdot Q_{tot,m}{}^{\alpha} = f^{\alpha} \cdot dH_{i,m}. \tag{9}$$

This equation shows that any multi-connected network has self-similarity properties between the total head losses in each node and an average total entering discharge. Then, for any deterministic scenario, the pressure head at each node is known from eqn (9) if the solution of the average demand scenario is known. Furthermore, eqn (7) shows that the deterministic scenarios are all represented by the reference DAG obtained for the average scenario.

3.3 Behaviour of WDNs in presence of stochastic water demand scenarios

The discussion just provided remains valid on average in the case that the demand scenarios are randomly perturbed with respect to the local average scenario of reference, respecting the scaling laws and the cross-correlation.

Each stochastic scenario can be compared with the average deterministic scenario obtained by considering the same total discharge value introduced into the network. These comparisons show that the differences in nodal demands between pairs of deterministic and stochastic scenarios are small, compared to the value of the demand.

This allows evaluating the trend of the nodal head even in non-deterministic scenarios. Moreover, due to the exiguous residual value, the formula proposed by Todini and Pilati [19] can be linearized as follows:

$$H_i = [A_{21}(D_{11})^{-1}A_{12}]^{-1}\{A_{21}(D_{11})^{-1}[(D_{11} - A_{11})]Q - A_{10}H_0 + q_i\}. \tag{10}$$

For this purpose, a first-order truncated Taylor expansion is performed on eqn (10):

$$\Delta dH_{i,r} = [A_{21} D_{11}^{-1} A_{12}]^{-1} \cdot \Delta q_{i,r}. \tag{11}$$

Consequently, the head loss of node i is given by the sum of the result obtained considering the related deterministic scenarios and the residual value of the head loss due to the stochastic scenario, obtained from the previous linearization.

$$dH_{i,r} = dH_{i,s} + \Delta dH_{i,r} = f^{\alpha} \cdot dH_{i,s} + \Delta dH_{i,r}. \tag{12}$$

The residuals obtained from the comparison between the local mean demands $q_{i,s}$ and the relative perturbed demand $q_{i,r}$ can be considered a random variable of assigned mean μ_{q_i} and variance $\sigma^2_{q_i}$. This random variable Δq is linked to the residual of the nodal head $\Delta dH_{i,r}$ by eqn (11). For the theory of random variables, the linear combination of random variables provides an expression of the expected value and variance for the derived variable. It is therefore possible to obtain the statistical parameters of the residual nodal heads by the demands':

$$E[\Delta dH_i] = [A_{21} D_{11}^{-1} A_{12}]^{-1} \cdot E[\Delta q_i]. \tag{13}$$

$$\sigma^2_{\Delta dHi} = [A_{21} D_{11}^{-1} A_{12}]^{-1} \cdot \{diag[\sigma^2_{\Delta dq_i}] + cov[\Delta Q_i, \Delta Q_j]\} \cdot \\ \cdot \{[A_{21} D_{11}^{-1} A_{12}]^{-1}\}^T \tag{14}$$

Considering eqns (13) and (14), it is possible to define the approximate values of the nodal head losses and of the pipe flow due to the random realization of a stochastic scenario by assigning only a total inflow. For stochastic water demand scenarios, the equations proposed by Todini and Pilati [19] results as:

$$H_{i,r} = H_0 - dH_{i,s} + [A_{21}(D_{11})^{-1}A_{12}]^{-1} \cdot E[\Delta q_i] + \{[A_{21}(D_{11})^{-1}A_{12}]^{-1}\}^2 \cdot \\ \cdot \sigma^2_{\Delta dHi} + 2 \cdot [A_{21}(D_{11})^{-1}A_{12}]^{-1} \cdot cov[\Delta Q_i, \Delta Q_j]. \tag{15}$$

$$Q_{k,r} = Q_{k,m} - (D_{11,m})^{-1}(A_{11,m} \cdot Q_{k,m} + A_{12} \cdot H_{i,r} + A_{10} \cdot H_0). \tag{16}$$

It can happen that in WDN with a non-redundant size the network DAG can undergo local reversals in some arcs. However, it is empirically demonstrated that the reference DAG remains the predominant DAG in analysed scenarios. Hence, the average scenario confirms its importance in defining the nodal behaviour, pointing out the filter effect of the network.

4 THEORICAL RESULT VALIDATION

The above theoretical analysis is verified in the follows through simulations on the WDN of Fossolo, a suburban area of Bologna, Italy (Fig. 1). The topology of this network was proposed by Bragalli et al. [22] and shared by the University of Exeter. The original WDN size has been modified so as to have a higher redundancy, as it occurs in real networks.

Water users are residential, and the annual average nodal demand is known. The number of users and the average peak hour demand in each node have been estimated using demand data by Bragalli et al. [22]. The sizing, the average annual demand and the estimated number of users are available in the appendix.

The demand factor and the correlation coefficient have been estimated through a preliminary analysis on historical data of a similar real residential dataset subjected to smart metering observation [23]. Hence, the statistical parameters for the generation of the stochastic scenarios were opportunely calibrated to comply with these values. The characteristics of the generated stochastic scenarios are presented in Table 1 and in Fig. 2.

The hydraulic simulation of the WDN allows obtaining the relation between nodal head loss dH_i and total inflow Q_{tot} (eqn (6)) in each node, both for the deterministic and stochastic scenarios. Fig. 3(a) reports the results for node (ID9): the red line refers to deterministic

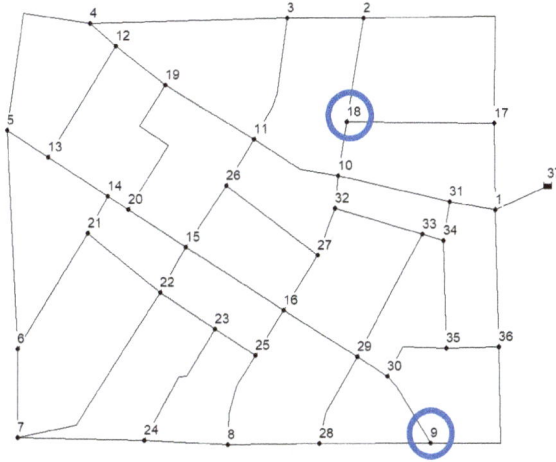

Figure 1: Fossolo network and node IDs.

Table 1: Statistical parameters used to generate stochastic scenarios.

Mean demand	Standard deviation	Cross-correlation
l/min	l/min	–
0.680	3.000	0.0016

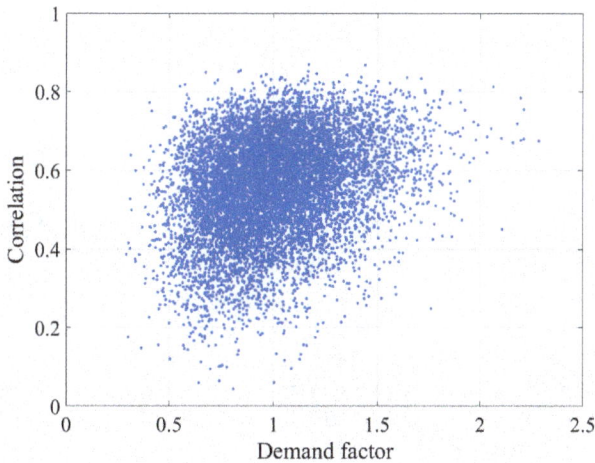

Figure 2: Demand factor and correlation of the generated stochastic scenarios.

scenarios, blue dots to stochastic scenarios. The red line of the deterministic scenarios is the mean curve of the stochastic scenarios. Fig. 3(b) reports the histogram of the residuals of the nodal head losses from the local mean value.

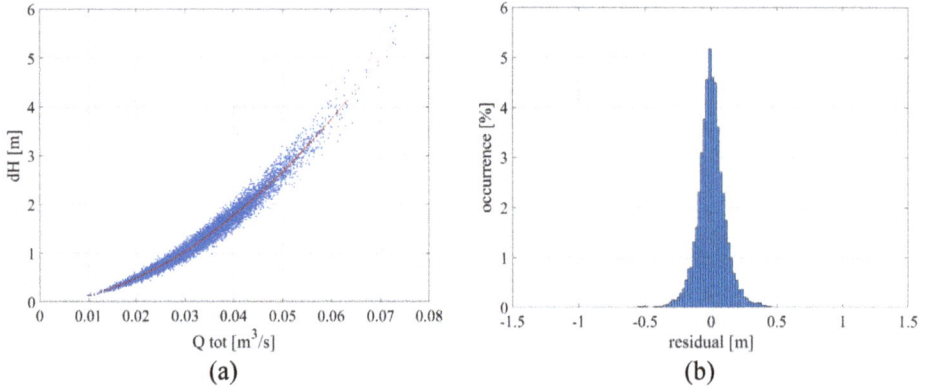

Figure 3:   (a) Nodal head losses of one node (ID9) of Fossolo WDN in case of deterministic scenarios (red line) and stochastic scenarios (blue points); and (b) histogram of the residuals of the nodal head losses from the local mean value (deterministic trend).

Fig. 4(a) and (b) reports comparable results for node (ID18), which is characterized by higher variability (highlighted by the two histograms in Fig. 3(b) and Fig. 4(b)). However, the residuals from the deterministic curve keep a normal type of probability distribution, even if the variance seems higher.

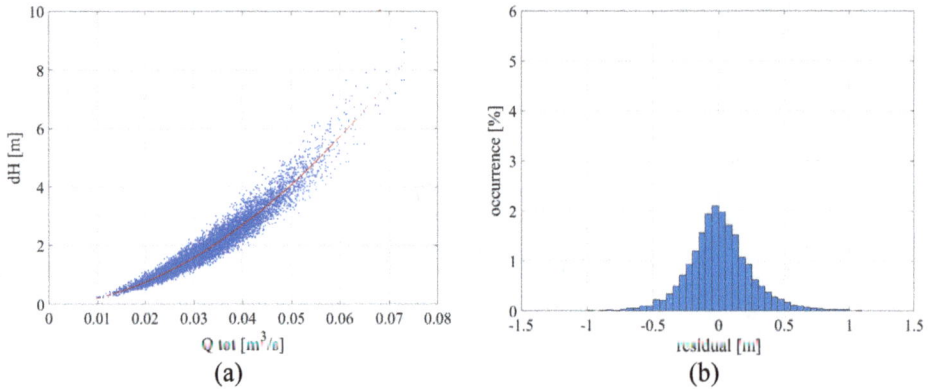

Figure 4:   (a) Nodal head losses of one node (ID18) of the Fossolo WDN in case of deterministic scenarios (red line) and stochastic scenarios (green points); and (b) histogram of the residuals of the nodal head losses from the local mean value (deterministic trend).

The residual values of the head losses of the single node allow calculating the eqn (14) and obtaining the estimated value of the variance of the residuals head losses with respect to the residual nodal demands.

Fig. 5 shows a comparison between the value assumed from the variance evaluated on the residuals obtained from the simulation of ten thousand stochastic scenarios, and the one estimated by solving the eqn (14). This figure shows the relation between the two variances with respect to each node, hence proving that the linearization provides with a sufficient degree of accuracy the differences within nodal head losses from their expected deterministic value.

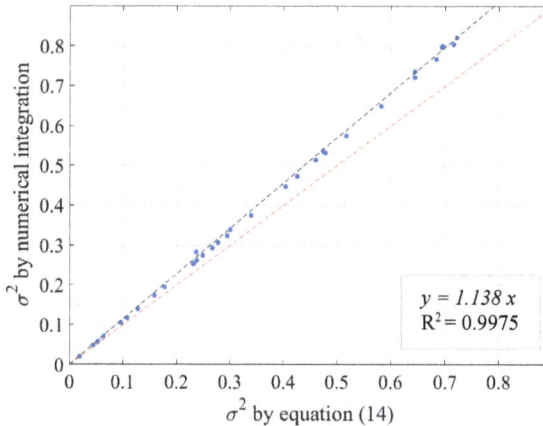

Figure 5:   Comparison between the nodal values of variance, that is the ones from linearization and those calculated for the 10,000 stochastic scenarios.

Fig. 5 shows a slight underestimation (13.8%) of the theoretical values compared to the empirical ones. This underestimation is due to the effect of linearization.

5 CONCLUSION

In this paper the hydraulic behaviour of the WDNs is presented as a function of the average demand scenario and network topology.

The theoretical formulation was validated considering both deterministic scenarios and stochastically generated stochastic scenarios.

The first finding is that considering any deterministic scenario, the pressure at each node can be evaluated if the pressure of an average demand scenario is known. Furthermore, all the deterministic scenarios can be represented by a reference DAG obtained considering the average scenario. This conclusion is also generally valid for randomly perturbed demand scenarios, as long as they respect the scaling laws and the cross-correlation, with respect to the reference average scenario.

By comparing deterministic and stochastic scenarios having the same total inletting discharge, we showed that the relative differences in nodal demands are small. This result allows evaluating the trend of the nodal head even in non-deterministic scenarios.

Furthermore, estimating pipe flow rates and nodal head losses can be considered as an optimal starting point for the resolution of the equation proposed by Todini and Pilati to increase the speed of convergence in the numerical solution of the hydraulic model.

From a practical point of view, the results achieved make it possible to provide an objective criterion for evaluating the pressures detected in each node based only on the overall flow rate value. This allows operational decisions to be made in real time.

NOTATION

The following symbols are used in the paper:

A_{10} = $[p, n_0]$ incidence matrix relating pipes to known head nodes;

A_{11} = $[p, p]$ diagonal matrix, whose generic term is $A_{11}(k, k) = r_k |Q_k|^{\alpha-1}$;

A_{12} = $A^T_{21} = [p, n]$ incidence matrix relating pipes to unknown head nodes;

$a_{i,j}$ = generic element of A_{12};

α = exponent. Its value is 1.852 when using the Hazen–Williams equation;

B = $[l, p]$ incidence matrix relating pipes to loops;

$\beta_{l,k}$ = generic element of B;

D_{11} = $[p, p]$ diagonal matrix of derivatives of A_{11} with respect to Q;

H_0 = $[n_0]$ length vector of known fixed head nodes;

i, j, k = generic index;

γ_{D_k} = $[p]$ vector of coefficient that depends by diameter and roughness of pipes;

l = number of independent loops;

L_k = $[p]$ vector of length of pipes;

n_i = number of pipes connected to node i;

n = number of nodes with unknown head;

n_0 = number of nodes with known head;

p = number of pipes;

Q = $[p]$ vector of pipe flows;

q = $[n]$ vector of nodal demands;

q_i = known nodal demand at node i;

$Q_{k_{i,j}}$ = flow in the generic pipe $k_{i,j}$ connected to node i;

Q_{tot} = total instantaneous discharge;

r_k = $[p]$ vector of coefficients that depends on the dimensions used on the pipe like diameter, roughness and length.

APPENDIX

Table A1: Average demand at peak hour in each node and relative number of users.

Node's ID	Peak nodal demand (l/s)	Number of users	Node's ID	Peak nodal demand (l/s)	Number of users	Node's ID	Peak nodal demand (l/s)	Number of users
1	0.49	126	13	1.16	297	25	0.77	198
2	1.04	267	14	0.54	141	26	1.69	435
3	1.02	261	15	1.10	282	27	1.42	366
4	0.81	210	16	1.21	312	28	0.30	78
5	0.63	162	17	1.27	327	29	0.62	159
6	0.79	204	18	2.02	519	30	0.54	141
7	0.26	69	19	1.88	483	31	0.90	231
8	0.58	150	20	0.93	240	32	1.03	264
9	0.54	141	21	0.96	246	33	0.77	198
10	1.11	285	22	0.97	249	34	0.74	192
11	1.75	450	23	0.86	222	35	1.16	297
12	0.91	234	24	0.67	174	36	0.47	123

Table A2: Pipes' size: Internal diameter (D) (mm), Length (L) (m), Roughness (C) ($mm^{1/3} \cdot s^{-1}$).

ID	D	L	C	ID	D	L	C	ID	D	L	C
1	144.8	132.8	130	21	113	84.0	140	41	45.2	203.8	150
2	67.8	374.7	130	22	126.6	49.8	130	42	45.2	248.1	150
3	45.2	119.7	140	23	144.8	78.5	130	43	45.2	65.2	145
4	67.8	312.7	140	24	81.4	99.3	140	44	57	210.1	145
5	45.2	289.1	150	25	99.4	82.3	140	45	67.8	147.6	145
6	67.8	336.3	145	26	57	147.5	140	46	67.8	103.8	140
7	67.8	135.8	145	27	67.8	197.3	140	47	45.2	211.0	140
8	57	201.3	145	28	113	83.3	140	48	81.4	75.1	140
9	57	132.5	145	29	45.2	113.8	140	49	113	180.3	150
10	45.2	144.7	145	30	81.4	80.8	140	50	57	149.1	140
11	45.2	175.7	145	31	45.2	341.0	140	51	57	215.1	130
12	81.4	112.2	145	32	57	77.4	150	52	81.4	144.4	130
13	99.4	210.7	130	33	45.2	112.4	150	53	99.4	34.7	130
14	162.8	75.4	130	34	45.2	37.3	145	54	126.6	59.9	130
15	99.4	181.4	130	35	45.2	108.9	145	55	67.8	165.7	130
16	67.8	147.0	150	36	81.4	182.8	150	56	99.4	120.0	130
17	81.4	162.7	150	37	113	136.0	140	57	67.8	83.2	130
18	45.2	99.6	150	38	99.4	56.7	140	58	203.4	1.0	130
19	57	53.0	150	39	81.4	124.1	130				
20	67.8	163.0	140	40	144.8	234.6	130				

REFERENCES

[1] Magini, R., Boniforti, M.A. & Guercio, R., Generating scenarios of cross-correlated demands for modelling water distribution networks. *Water*, **11**(493), 2019.

[2] Creaco, E., Galuppini, G., Campisano, A. & Franchini, M., Bottom-up generation of peak demand scenarios in water distribution networks. *Sustainability*, **13**(31), 2021.

[3] Berardi, L. & Giustolisi, O., Calibration of design models for leakage management of water distribution networks. *Water Resources Management*, **35**(2), 2021.

[4] Wéber, R. & Hős, C., Efficient technique for pipe roughness calibration and sensor placement for water distribution systems. *Journal of Water Resources Management*, **146**(1), 2020.

[5] Benefield, L.A., Rule development committee issue research report draft: Residential flow rates, Department of Health, Washington State, 2002.

[6] Director of Santa Rosa Water/City Engineer, Guide to potable water, recycled water and wastewater policy, City of Santa Rosa, CA, 2022.

[7] Department of Water and Waste, Wastewater flow estimation and servicing guidelines, City of Winnipeg, Manitoba, 2020.

[8] Chu, S. et al., Dealing with data missing and outlier to calibrate nodal water demands in water distribution systems. *Water Resources Management*, **35**(9), pp. 2863–2878, 2021.

[9] Meirelles, G. et al., Calibration model for water distribution network using pressures estimated by artificial neural networks. *Water Resources Management*, **34**(1), pp. 4339–4351, 2017.

[10] Kapelan, Z., Savic, D. & Walters, G., Calibration of water distribution hydraulic models using a Bayesian-type procedure. *Journal of Hydraulic Engineering*, **133**(8), pp. 927–936, 2007.

[11] Magini, R., Pallavicini, I. & Guercio, R., Spatial and temporal scaling properties of water demand. *Journal of Water Resources Planning and Management*, **134**, pp. 276–284, 2008.

[12] McKay, M., Conover, W. & Beckman, R., A comparison of three methods for selecting values of input variables in the analysis of output from a computer code. *Technometrics*, **21**, pp. 239–245, 1979.

[13] Cario, M.C. & Nelson, B.L., Modeling and generating random vectors with arbitrary marginal distributions and correlation matrix. Department of Industrial Engineering and Management Sciences, Northwestern University, Evanston, IL, 1997.

[14] Iman, R.L. & Conover, W.J., A distribution-free approach to inducing rank correlation among input variables. *Communications in Statistics*, **11**(3), pp. 311–334, 1982.

[15] Russo Spena, A., Su alcune proprietà caratteristiche delle reti, Napoli, 1950.

[16] Russo Spena, F. & Vacca, A., Una formulazione variazionale complementare del problema di verifica di reti di distribuzione di fluiso incompressibile. *Ratio Mathematica – Journal of Mathematics, Statistics, and Applications*, **4**, 1992.

[17] Cross, H., Analysis of flow in networks of conduits or conductors. *Bulletin No. 286, Engineering Experiment Station*, University of Illinois, 1936.

[18] Wood, D.J. & Charles, C.O.A., Hydraulic network analysis using linear theory. *Journal of the Hydraulic Division*, **98**(HY7), pp. 1157–1170, 1972.

[19] Todini, E. & Pilati, S., A gradient algorithm for the analysis of pipe networks. *Computer Application in Water Supply*, **1**, pp. 1–20, 1989.

[20] Todini, E., A more realistic approach to the extended period simulation of water distribution networks. *Advanced in Water Supply Management*, pp. 173–183, 2003.

[21] Todini, E. & Rossman, L., Unified framework for deriving simultaneous equation algorithms for water distribution networks. *Journal of Hydraulic Engineer*, pp. 511–526, 2013.

[22] Bragalli, C., D'Ambrosio, C., Lee, J., Lodi, A. & Toth, P., Water network design by MINLP. Report No. RC24495. Tech. rep., IBM Research Report, 2008.

[23] Del Giudice, G., Di Cristo, C. & Padulano, R. Spatial aggregation effect on water demand peak factor. *Water*, **12**(7), 2020.

PRESSURES–IMPACT ANALYSIS OF THE PACORA RIVER, PANAMA

QUIRIATJARYN M. ORTEGA-SAMANIEGO[1,2], ANDRES FRAIZ[3], ARTURO DOMINICI[4],
ADRIAN RAMOS-MERCHANTE[5], MARÍA PACHES[1] & INMACULADA ROMERO[1]
[1]Research Institute of Water and Environmental Engineering, (IIAMA), Universitat Politècnica de València, Spain
[2]Ministry of Environment of Panama, Panama
[3]Wetlands International, Panama
[4]International Maritime University of Panama, Panama
[5]University of Huelva, Spain

ABSTRACT

The Pacora River is one of the priority basins of the Republic of Panama since it provides water for human consumption therefore its conservation is of great importance; the country has laws that protect the integrity of the basins from their ecological nature to that for use for economic activities. This basin has great potential not only for agricultural activities but also a buffer for population development in the urban area of the country. This study was carried out in the middle and lower upper parts of the Pacora River during the wet, dry and transitional season, the data obtained was used to calculate the water quality index and the fluvial habitat index and the elaboration of an impact–pressure matrix, in order to evaluate the levels of influence that anthropogenic activities have in the area, based on established laws regulating water quality levels and forest legislation. The results obtained indicate that the Pacora River are among the ranges of little polluted, acceptable and uncontaminated for the water quality index and for the habitat fluvial index in the range of not reaching very good, good and very good. Regarding the analysis of pressures and impacts, the values are between probable impact, checked and non-apparent impact and medium to high risk. These data indicate a shift in the use of native forest land to industrial, agricultural and urban activities.

Keywords: pressures–impacts analysis, water quality index (WQI), fluvial habitat index (FHI), Pacora River, Panama.

1 INTRODUCTION

Water is a liquid of vital importance for the maintenance of life in the environment, although the quality of water has deteriorated over time, both in surface and groundwater, due to natural and human activities [1]. Atmospheric, hydrological, climatic, lithological and topographical factors are mentioned as natural factors that interfere with the quality of the waters [2], on the other hand, the activities of anthropogenic origin that go from the rural landscape to the urban, such as livestock, agriculture, mining, industries, residential areas and the disposal of liquids such as solids, in addition to the loss of green areas replaced by cement, increase runoff, considering that problems of sedimentation, erosion, presence of heavy metals, fecal coliforms, agrochemicals, among others [3], [4], and contamination limits the use of water for recreational activities [5]. In addition, it must be considered that the pollution that arrives through the waters that flow through rivers and streams reaches the sea (it has been estimated that 90% of these pollutants are transported to the sea via rivers), entailing a series of consequences negative effects on estuaries and on the other living resources that inhabit these ecosystems [6]. According to UNEP [7], water pollution has worsened since the 1990s in most rivers in Latin America. Global concern about water availability and quality has been increasing, with demand for water estimated to increase by 20–30% by 2050 [8].

A successful tool to verify and determine contamination is through the water quality index (WQI), which represents a valuable and unique qualification to show the general state of

WIT Transactions on The Built Environment, Vol 208, © 2022 WIT Press
www.witpress.com, ISSN 1743-3509 (on-line)
doi:10.2495/FRIAR220101

water quality that is useful for the selection of the technique that mitigates with the use of an appropriate treatment to resolve or improve water quality [9], [10].

2 MATERIALS AND METHODS

2.1 Study area

The Pacora River is located on the Pacific slope of the Republic of Panama at the coordinates 8°00' and 8°20' north latitude and 79°15' and 79°30' west longitude. Its extension is 388 km² and the drainage area is 364 km². It is bordered on the north by the Chagres River, on the south by the Bay of Panama, on the east by the Bayano River and on the west by the Juan Diaz River. The basin is represented in three life zones: very humid premontane forest, premontane humid forest and tropical humid forest. This river provides water for human consumption and therefore is considered one of the priority basins in the country [11].

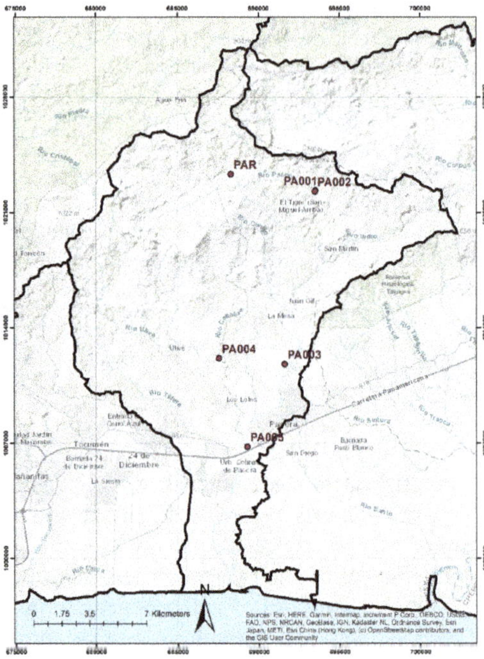

Figure 1: Location of the study sites.

2.2 Pressure impact analysis

For the development of this analysis, an adaptation was made to the manual prepared by the General Directorate of Water of the Ministry of the Environment of Spain IMPRESS (2004), whose objective is to identify the risks of surface water bodies for compliance with the Water Framework Directive 2000/60/EC [12].

The pressure–impact assessment was carried out at a qualitative level considering the results obtained from the water quality index and the fluvial habitat index, with these data it was classified in impacts: probable, checked and no apparent impact, and according to risk:

high, medium, low and non-risk, based on established laws regulating water quality levels and forest legislation.

3 RESULTS AND DISCUSSION

3.1 Pressure impact analysis

3.1.1 PA001
The point PA001WE had a value of 76 and 83 for PA001DR in the water quality index whose rating is acceptable, however, these scores decrease in the parameters established in Executive Decree No. 75 (of June 4, 2008) "By which the primary standard of environmental quality and quality levels for inland waters for recreational use with and without direct contact is dictated", in its Article 12 [13]. The point PA001TR with a value of 92 whose rating is uncontaminated; these values are due to the population settlements in the area.

Regarding the river habitat index its value was 72 whose rating is very good however its score decreases in the percentage of shade in the channel therefore does not fully comply with Law 1 of February 3, 1994, Chapter III of the Forest Protection "In the rivers and streams the width of the channel will be taken into consideration and the width of it will be left to both sides but in no case will it be less than (10) meters; a strip of forest of not less than ten (10) meters may also be left as a distance" [14]. Therefore, PA001 has a probable impact and a medium risk according to the values of WQI and HFI.

3.1.2 PA002
The point PA002WE had a value of 74, PA002DR 75 and PA002 TR 75 in the water quality index whose rating is acceptable, however, these scores decrease in the parameters established in Executive Decree No. 75 (of June 4, 2008) "By which the primary standard of environmental quality and quality levels for inland waters for recreational use with and without direct contact is dictated", in its Article 12 [13]. The point, these values are due to the population settlements in the area.

Regarding the river habitat index its value was 58 for the points PA002WE, PA002DR AND PA002TR whose rating is good however its score decreases in the percentage of shade in the channel therefore does not fully comply with Law 1 of February 3, 1994, Chapter III of the Forest Protection "In rivers and streams, the width of the channel will be taken into consideration and the width of the same will be left on both sides but in no case will it be less than (10) meters; a strip of forest of not less than ten (10) meters may also be left as a distance" [14]. Therefore, PA002 has a probable impact and a medium risk according to the values of WQI and HFI.

3.1.3 PA003
The point PA003WE had a value of 73, PA003DR 77 and PA003TR 73 in the water quality index whose rating is acceptable, however, these scores decrease in the parameters established in Executive Decree No. 75 (of June 4, 2008) "By which the primary standard of environmental quality and quality levels for inland waters for recreational use with and without direct contact is dictated", in its Article 12 [13]. The point, these values are due to the population settlements in the area.

Regarding the river habitat index its value was 54 for the points PA003WE, PA003DR and 67 PA003TR whose rating is good however its score decreases in the percentage of shade in the channel therefore does not fully comply with Law 1 of February 3, 1994, Chapter III of the Forest Protection "In rivers and streams, the width of the channel will be taken into consideration and the width of the same will be left on both sides but in no case will it be less than (10) meters; a strip of forest of not less than ten (10) meters may also be left as a distance" [14]. Therefore, PA003 has a likely impact and a high risk according to the values of WQI and HFI.

3.1.4 PA004

The point PA004WE had a value of 73, PA004DR 79 and PA004TR 77 in the water quality index whose rating is acceptable, however, these scores decrease in the parameters established in Executive Decree No. 75 (of June 4, 2008) "By which the primary standard of environmental quality and quality levels for inland waters for recreational use with and without direct contact is dictated", in its Article 12 [13]. The point, these values are due to the population settlements in the area.

Regarding the river habitat index its value was 61 for points PA004WE, PA004DR and 64 PA004TR whose rating is good however its score decreases in the percentage of shade in the channel therefore does not fully comply with Law 1 of February 3, 1994, Chapter III of the Forest Protection "In rivers and streams, the width of the channel will be taken into consideration and the width of the same will be left on both sides but in no case will it be less than (10) meters; a strip of forest of not less than ten (10) meters may also be left as a distance" [14]. Therefore, PA004 has a likely impact and a high risk according to the values of WQI and

3.1.5 PA005

The point PA005WE had a value of 68, rating of little contaminated, PA005DR 78 and PA005TR 76 is acceptable, in the water quality index, however, these scores decrease in the parameters established in Executive Decree No. 75 (of June 4, 2008) "By which the primary standard of environmental quality and quality levels for inland waters of recreational use with and without direct contact is dictated", in its Article 12 [13]. The point, these values are due to the population settlements in the area.

Regarding the river habitat index, its value was 50 for points PA005WE and PA005DR whose rating does not reach good and 57 PA005TR whose rating is good however its score decreases in the percentage of shade in the channel therefore does not fully comply with Law 1 of February 3, 1994, Chapter III of the Forest Protection "In the rivers and streams the width of the channel will be taken into consideration and the width of the same will be left on both sides but in no case will it be less than (10) meters; a strip of forest of not less than ten (10) meters may also be left as a distance" [14]. Therefore, PA005 has a checked impact and a high risk according to the values of WQI and FHI.

3.1.6 PAR

This point is the reference site, represents the ideal conditions, had a value of 91, rating of uncontaminated, in the Water Quality index, however, these scores decrease in the parameters established in Executive Decree No. 75 (of June 4, 2008) "By which the primary standard of environmental quality and quality levels for inland waters for recreational use with and without direct contact is dictated", in its Article 12 [13]. The point, these values are due to the population settlements in the area.

Regarding the river habitat index its value was 86 whose rating is very good however its score decreases in the percentage of shade in the channel therefore does not fully comply with Law 1 of February 3, 1994, Chapter III of the Forest Protection "In the rivers and streams the width of the channel will be taken into consideration and the width of it will be left on both sides but in no case will it be less than (10) meters; a strip of forest of not less than ten

(10) meters may also be left as a distance" [14]. Therefore, PA005 has a non-apparent impact and an average risk according to the values of WQI and FHI.

Table 1:   WQI and FHI ratings of the monitored sites and impact–pressure matrix of the Pacora River.

COD	WQI	HFI	Impact	Risk	Site
PA001WE	76	72	Probable	Middle	
PA001DR	83	72	Probable	Middle	
PA001TR	92	72	Probable	Middle	
PA002WE	74	58	Probable	Middle	
PA002DR	75	58	Probable	Middle	
PA002TR	75	58	Probable	Middle	
PA003WE	73	54	Probable	High	
PA003DR	77	54	Probable	High	
PA003TR	73	67	Probable	High	
PA004WE	73	61	Probable	High	
PA004DR	79	61	Probable	High	
PA004TR	77	54	Probable	High	
PA005WE	68	50	Checked	High	
PA005DR	78	50	Checked	High	
PA005TR	76	57	Checked	High	
PAR	91	86	No apparent impact	Middle	

The Pacora River has experienced a population growth in its middle and lower parts, therefore a change of land focused on industrial and urban development. In the upper parts there is a change of native forest species due to the development of agricultural activity and also the extraction of minerals directly from the river [15].

The water quality index of the Pacora River is in the range of acceptable, suitable for recreational, industrial, agricultural activities, aquatic biodiversity and human consumption [11], [16].

Different authors agree that the review of water quality data over time is the best way to verify the behavior of surface water pollution that passes through different areas between agricultural, urban, whether residential or industrial, such as the case of the Yamuna River in India between the years 2000 and 2009 [17]; Juan Díaz River in Panama between the years 2002 and 2018 [18]; comparative studies between water bodies in the USA and Canada [19]; monitoring of the four main rivers of South Korea between the years 2007 and 2018 [20]. Although the results present a variability in the results depending on the location of the sampling station, in this sense the concept of water quality is not necessarily a state of purity of the water, but refers to the chemical, physical and biological characteristics that determine their different uses [21].

4 CONCLUSION

The results obtained in this research indicate that the Pacora River has a score of 73 to 92 with ratings of little contaminated, acceptable and uncontaminated in the Water Quality Index. For the river habitat index we find values from 50 to 86 with ratings of not reaching good, good and very good

The Pacora River for points PA001, PA002 has a probable impact and medium risk, for points PA003 and PA004 a probable impact and high risk, for point PA005 a checked impact and high irrigation and finally PAR the reference point non-apparent impact and medium risk.

These results show the change of native forest soil to the development of industrial, agricultural and urban activities.

ACKNOWLEDGEMENTS

This research was financed by the Scholarship of the Subprogram of Doctoral and Postdoctoral Scholarships of the National Secretariat of Science and Technology (SENACYT) in conjunction with the Institute for the Training and Use of Human Resources (IFARHU). This research is part of the project Environmental Impact of Multiple Stressors in Aquatic Ecosystems of the Metropolitan Area of Panama, financed by SENACYT.

REFERENCES

[1] Uddin, M.G., Nash, S. & Olbert, A., A review of water quality index models and their use for assessing surface water quality. *Ecological Indicators*, **122**, 2021.
[2] Uddin, M.G., Moniruzzaman, M., Quader, M.A. & Hasan, M.A., Spatial variability in the distribution of trace metals in groundwater around the Rooppur nuclear power plant in Ishwardi, Bangladesh. *Groundw. Sustain. Dev.*, 2018.
[3] Sánchez, E., Colmenarejo, M.F., Vicente, J., Rubio, A., García, M.G., Travieso, L. & Borja, R. Use of the water quality index and dissolved oxygen deficit as simple indicators of watersheds pollution. *Ecological Indicators,* **7**, pp. 315–328, 2007.

[4]   Lobato, T.C., Hauser, R.A., David, T.C., Oliveira, T.F., Silveira, A.M., Silva, H.A.N., Tavares, M.R.M. & Saraiva, A.C.F., Construction of a novel water quality index and quality indicator for reservoir water quality evaluation: A case study in the Amazon region. *Journal of Hydrology*, **522**, pp. 674–683, 2015.

[5]   Lopes, F.W.A., Davies-Colley, R.J., Von Sperling, E. & Magalhães, Jr., A water quality index for recreation in Brazilian freshwaters. *Journal of Water and Health*, **14**(2), pp. 243–254, 2016.

[6]   Escobar, J., Perfil de la Contaminación de las Aguas Continentales Panameñas. Comisión de Pesca Continental para América Latina. FAO/COPESCAL, 1983.

[7]   UNEP (United Nations Environment Programme), A snapshot of the world's water quality: Towards a global assessment, 2016.

[8]   WWAP, *The United Nations World Water Development Report 2019: Leaving No One Behind*, 2019.

[9]   Tyagi, S., Sharma, B., Singh, P. & Dobhal, R., Water quality assessment in terms of water quality index. *Am. J. Water Resour.*, **1**(3), pp. 34–38, 2013.

[10]   Sha, K.A. & Joshi, G., Evaluation of water quality index for River Sabarmati, Gujarat, India. *Applied Water Science*, **7**, pp. 1349–1358, 2017.

[11]   ANAM (National Environmental Authority), Report on the monitoring of water quality in the watersheds of Panama, 2009–2012.

[12]   Dirección General del Agua del Ministerio de Medio Ambiente, Manual para el análisis de presiones e impactos relacionados con la contaminación de las masas de agua superficiales. España, p. 49, 2004.

[13]   Ministerio de Economía y Finanzas Decreto Ejecutivo No. 75 (de 4 de junio de 2008) Por el cual se dicta la norma primaria de calidad ambiental y niveles de calidad para las aguas continentales de uso recreativo con y sin contacto directo, República de Panamá.

[14]   Autoridad Nacional del Ambiente Legislación Forestal de la República de Panamá Ley No 1 Por La Cual Se Establece La Legislación Forestal En La República De Panamá y Se Dictan Otras Disposiciones, República de Panamá.

[15]   Programa Regional para la Reducción de la Vulnerabilidad y Degradación Ambiental. Formulación del Plan Estratégico para el Manejo Integrado de la Cuenca del Río Pacora. Panamá, República de Panamá, p. 121, 2008.

[16]   Cornejo, A. et al., Diagnosis of the environmental condition of the superficial tributaries of Panama, 2017.

[17]   Sharma, D. & Kansal, A., Water quality analysis of River Yamuna using water quality index in the national capital territory, India (2000–2009). *Appl. Water Sci.*, **1**, pp. 147–157, 2011.

[18]   Ortega-Samaniego, Q., Romero, I., Paches, M., Dominici, A. & Fraiz, A., Assessment of physicochemical and bacteriological parameters in the surface water of the Juan Díaz River, Panamá. *WIT Transactions on Ecology and the Environment*, vol. 251, WIT Press: Southampton and Boston, 2021.

[19]   Lumb, A., Sharma, T.C. & Bideault, J.-F., A comparative study of USA and Canadian water quality index models. W*ater Quality, Exposure and Health*, **3**, pp 203–216, 2012.

[20]   Kim, T., Kim, Y., Shin, J., Go, B. & Cha, Y., Assessing land-cover effects on stream water quality in metropolitan areas using the water quality index. *Water*, **12**(11), p. 3294, 2020.

[21] Merten, G.H. & Minella, J.P., Qualidade da água em bacias hidrográficas rurais: um desafio atual para sobrevivência futura. *Agroecologia e Desenvolvimento Rural Sustentável*, **3**(4), pp. 33–38, 2002.

Author index

WITPRESS   ...for scientists by scientists

Air and Water Pollution XXX

*Edited by: **S. MAMBRETTI**, Polytechnic of Milan, Italy, **J. LONGHURST**, University of the West of England, UK and **J. BARNES**, University of the West of England, UK*

The merger of two successful events to form the 30th International Conference on Modelling, Monitoring and Management of Air and Water Pollution provided the papers that are published in this volume.

Many important air pollution issues are discussed, demonstrating the widespread nature of the air pollution phenomena and the in-depth exploration required to address their impacts on human health and the environment.

The scientific knowledge derived from well-designed studies needs to be allied with further technical and economic studies in order to ensure cost-effective and efficient mitigation. In turn, the science, technology and economic outcomes are necessary but not sufficient. The outcome of such research needs to be contextualised within well-formulated communication strategies that help policymakers and citizens to understand and appreciate the risks and rewards arising from air pollution management.

In addition, the topic of Water Pollution is discussed in a number of contexts across different areas of water contamination.

The environmental problems caused by the increase of pollutant loads discharged into natural water bodies requires the formation of a framework for regulation and control. This framework needs to be based on scientific results that relate pollutant discharge with changes in water quality. The results of these studies allow the industry to apply more efficient methods of controlling and treating waste loads, and water authorities to enforce appropriate regulations regarding this matter.

Environmental problems are essentially interdisciplinary. Engineers and scientists working in this field must be familiar with a wide range of issues including the physical processes of mixing and dilution, chemical and biological processes, mathematical modelling, data acquisition and measurement, to name but a few. In view of the scarcity of available data, it is important that experiences are shared on an international basis. Thus, a continuous exchange of information between scientists from different countries is essential.

WIT Transactions on the Ecology and the Environment, vol. 259

ISBN: 978-1-78466-467-1   eISBN: 978-1-78466-468-8
ISSN (print): 1746-4498   ISSN (online): 1743-3509

Published 2022 / 174pp

www.ingramcontent.com/pod-product-compliance
Lightning Source LLC
Chambersburg PA
CBHW081542220326
41598CB00036B/6528